How Was Our Universe Created?

How Was Our Universe Created ?

Adrian Bjornson

Addison Press
Woburn, Massachusetts
www.olduniverse.com

Published by: **Addison Press**
400 West Cummings Park
PMB 1725-111
Woburn, MA 01801
www.olduniverse.com

Copyright © 2001 by Addison Press

All rights reserved under International and Pan-American Copyright Conventions. No part of this book may be reproduced by any mechanical, photographic, or electronic process, nor may it be stored in a retrieval system, transmitted, or otherwise copied for public or private use, without written permission of Addison Press.

Printed in the United States of America

Publisher's Cataloguing-in-Publication data
    (Provided by Quality books, Inc.)
Bjornson, Adrian
    How Was Our Universe Created? / Adrian Bjornson. - -
1st ed.
    p. cm.
    Includes bibliographical references and index.
    ISBN 09703231-1-5
    1. Cosmology--Popular works. 2. Gravitation.
    3. Astrophysics. 4. Relativity (Physics) 5. Yilmaz,
    Huseyin. 6.Einstein, Albert 1879-1955. I. Title.
    QB982.B56 2001   523.1
                    QB101-200232

This book is dedicated to

Professor Huseyin Yilmaz

*Who has devoted his life to develop his theory of gravity,
which extends the Einstein relativity concepts
to achieve the goal that Albert Einstein sought.*

Acknowledgements

I am grateful for the patient assistance that Prof. Huseyin Yilmaz has provided in explaining his *Theory of Gravity* and the principles of the Einstein *General Theory of Relativity*.

I thank William C. Keel of the University of Alabama for his excellent photograph of the M51 Whirlpool Galaxy with its companion galaxy NGC 5195. This was taken on the 1.1-meter Hall telescope at the Lowell Observatory. I have used this photograph on the front cover and in Figure 1-1.

Foreword

This book explores in depth the mysteries of our universe in a manner that can be easily comprehended by a broad audience. This is a difficult requirement, because we must understand the Einstein general theory of relativity in order to investigate gravitational effects at cosmological distances. We must also understand the gravitational theory developed by Prof. Huseyin Yilmaz, which is a refinement of the Einstein theory. As we will see, the Yilmaz theory has corrected weaknesses in the Einstein theory that have resulted in non-physical cosmological predictions.

To satisfy our requirement, this book must be able to explain the Einstein and Yilmaz theories in such a manner that they can be easily understood by the general public. Some of the material presented is controversial, and therefore requires detailed substantiation to be accepted. Consequently our simple explanations cannot be superficial. These formidable requirements are satisfied by utilizing a series of documents having different levels of difficulty.

The material of this book is supported by the following reference, which has a general but more restricted audience than this book:

[1] Adrian Bjornson, *A Universe that We Can Believe*, Addison Press, Woburn, MA, 2000, ISBN 09703231-0-7 .

In the present book, the above document is called *Universe* [1]. It has two distinct parts: (1) the body, and (2) the appendices. The body stands by itself and can be understood without reading the appendices. The appendices give mathematical analyses that support the material in the body of the book. Nearly all of the material in the appendices can be followed by a person with a high-school background in algebra and geometry. The body should be understandable to the average reader, but has more technical detail than this present book.

The material in *Universe* [1] in turn is supported by theoretical analyses in the following document, which is available at no cost on the Internet:

[2] Adrian Bjornson, *Addendum to A Universe that We Can Believe"*, available on Internet Website *www.olduniverse.com.*

In the present book, the above document is called *Addendum* [2]. This document assumes the knowledge of calculus, and is directed toward a reader with scientific training. It gives detailed analyses related to the Einstein and Yilmaz theories, but does not require any specialized knowledge of relativity theory beyond what is presented in Universe [1]. It is recommended that one read *Universe* [1] before the website *Addendum* [2].

This book and Universe [1] can be purchased on the website *www.olduniverse.com*, which is specified in the Bibliography as reference [3]. This website also gives a summary of *Universe* [1].

This hierarchy of documents allows this book to pursue in a simple yet in-depth manner the difficult theoretical issues of cosmology. Detailed aspects of the discussion are relegated to *Universe* [1], which can be read by those who desire further support of the concepts. Readers with calculus background can investigate the more theoretical analyses that are presented in *Addendum* [2]. An individual with scientific training will realize that the concepts that are presented very simply in this book have a solid scientific foundation.

In this manner, we have the background to explore the mysteries of cosmology. We will find that the book presents an exciting new concept concerning our universe. Is it correct? It certainly rests on much firmer scientific ground than any other explanation of our universe.

Contents

Dedication and acknowledgements	v
Foreword	vii
Preface	xiii
1. Introduction	**1**
Our beautiful and mystifying universe	1
The Einstein gravitational theory	2
The big bang explanation of our universe	3
Need for a reliable theory of gravity	4
The Yilmaz cosmology model	5
Our picture of the universe	5
2. The big bang theory	**7**
The big bang and black hole singularities	7
Measuring the Hubble expansion of the universe	11
Development of the big bang theory	15
The apparent age of the universe	*15*
Historical development of the big bang theory	*15*
Effect of the computer on cosmological studies	*17*
Cosmic microwave background radiation	*20*
Evidence against the big bang	22
The editorial by Professor Geoffrey Burbidge	*23*
The big bang age dilemma	*24*
Mythological philosophy of big bang research	*26*
Quasar studies by astronomer Halton Arp	*29*
Lack of scientific objectivity in astronomy today	*32*
3. Alternatives to explain the Hubble redshift	**33**
Well-known alternatives to the big bang	33
Steady-state universe theory	*33*
Quasi steady-state cosmology model	*35*
Variable mass theory	*36*
Matter-antimatter theory	*37*
Photon collisions with hydrogen molecules	*37*
Other hypotheses for explaining the Hubble redshift	*39*
Summary of well-known alternatives to the big bang	*39*
The Yilmaz gravitational theory	39
Development of the Yilmaz theory of gravity	*39*
Simplicity of the Yilmaz theory	*40*
Failure of the Einstein theory in cosmology	*41*
The Yilmaz cosmology model	42
Basic predictions of the cosmology model	*42*
Cosmic microwave blackbody radiation	*43*
Picture of the universe derived from Yilmaz cosmology model	*43*

4. How was our world created?   45

The development of life on earth   45
Early life   *45*
The first animals   *47*
Development of fishes   *48*
Amphibians invade the land   *48*
Spread of plants over the land   *49*
The reign of the reptiles   *50*
The slow rise of the mammals   *52*
The ascent of humans   *54*
The creation of our solar system   55
The Hubble expansion of our universe   57

5. Discovery of the nature of light   59

Early concepts of light   59
Galileo, Copernicus, and Kepler   60
The discoveries of Isaac Newton   62
The wave nature of light   67
Electromagnetic wave concept   70
Search for the velocity of the luminiferous aether   74
Was the wave theory of light correct?   76

6. The Einstein special theory of relativity   77

Measuring the speed of sound   77
Measuring the speed of light   79
The Einstein theory of relativity   82
Implications of relativistic effects   84
Equivalence between energy and matter   84
The principle of covariance   85

7. General relativity   87

Generalizing the relativity principle   87
Applying the equivalence principle   89
The meaning of a tensor   93
Specifying a vector   *93*
Tensor to specify stress within a body   *97*
Tensors in relativity theory   *100*
The metric tensor   *101*
Relativistic effects produced by gravity   102
Formulas for relativistic effects   *102*
Plots of gravitational effects   *104*

8. The black hole and the quasar   109

The physically impossible black hole   110
The quasar enigma   113
Cause of the quasar redshift   *113*
Quasar observations of Halton Arp   *114*
Galaxies with intrinsic redshift   *116*
Other quasar enigmas   *117*

9. Application of the Einstein and Yilmaz theories — 119
The Einstein gravitational field equation — 119
The Yilmaz gravitational theory — 122
The gravitational field equations for the Einstein and Yilmaz theories — 124
Uniqueness of Yilmaz solution — 126
How Yilmaz derived his theory — 127

10. Weaknesses of the Einstein theory — 128
Does not achieve a two-body solution — 128
Einstein curvature tensor in a vacuum — 130
Conservation of matter-plus-energy — 132
Singularities predicted by the Einstein theory — 133
Consistency with quantum mechanics — 134

11. The Yilmaz cosmology model — 135
Description of Yilmaz cosmology model — 135
Reduction of speed of light, clock rate, and spatial dimensions — *135*
The Hubble expansion of the universe — *139*
How can gravity make the universe expand? — *142*
Creation of matter — *143*
Cosmic microwave background radiation — 143
Density of matter in the universe — 148
Uniqueness of cosmology model predictions — 149

12. Conclusions — 150
The implications of the Yilmaz cosmology model — 150
Our picture of the universe — *150*
The second law of thermodynamics — *155*
The size of the universe — *155*
Mythological status of astronomy today — 157
The plasma research of Hannes Alfven and Eric Lerner — *157*
Observations of quasar redshifts by astronomer Halton Arp — *159*
Contradiction with the philosophy of Albert Einstein — *160*
Need for a theory of gravity that works — 161
Religious and philosophical implications of our picture of the universe — 162

Bibliography — 165
Index — 167-173

Preface

The Creation of our Universe

Since the beginning of historical time, humans have asked: "How were we created?", and, "How was our world created?"

Scientists have come to understand, at least in a general sense, how our earth was formed, how life has evolved on earth, and how mankind has developed. They have also learned a great deal about how our sun was formed, along with its solar system that includes our earth. These concepts are discussed in this book to give the reader a broad view of the fundamental basis for our existence. On the other hand, the evolution of our universe remains a deep mystery. This book focuses on that great question, "How Was Our Universe Created?"

In recent years, many scientists have become convinced that they understand the manner in which our universe evolved. They claim that our universe suddenly began as an enormously dense mass that exploded with a "Big Bang" about 15 billion years ago. Nevertheless, there is a great deal of scientific evidence contradicting this theory. Despite the strong endorsement of the Big Bang concept by the astronomical community, it is highly speculative.

There are many variations of the Big Bang theory, which have been developed to explain its conflicts with observational evidence. Many alternatives to the Big Bang theory have also been proposed. However, all of these well-known cosmology theories have serious weaknesses, and so they all must be regarded as speculative.

The Einstein General Theory of Relativity

As we examine these diverse cosmology theories, we find that there is great similarity among them. Practically all of the theories have the same mathematical foundation: the Einstein general theory of relativity.

Therefore, in order for us to study cosmology we must understand the Einstein theory.

It is commonly believed that the Einstein general theory of relativity is so complicated it is beyond the comprehension of ordinary humans. There is a widely held myth that only an elite group of highly trained scientists can fathom the deep mysteries of the Einstein theory. This book explodes that myth. Although the detailed calculations of general relativity are very complicated, the principles involved in Einstein's mysterious equations can be explained in a simple manner that is understandable to the average reader. This philosophy will be demonstrated as the reader travels through the book.

The Einstein general theory of relativity was a monumental advance in our knowledge of physics. It contains many fundamental principles that are very sound. This theory is specified by a tensor equation called the Einstein gravitational field equation, which actually represents a set of 10 independent equations. This tensor equation is so complicated it could only be solved for very simple cases during Einstein's lifetime. It was not until a decade after Einstein's death, when powerful computers became available, that the Einstein gravitational field equation could be applied to general physical models.

Scientists have long questioned the adequacy of the Einstein gravitational field equation, because Einstein derived this equation in a somewhat intuitive manner. Over the years, various alternatives to it have been proposed by responsible scientists.

Starting in the 1960's, hundreds of theoreticians have developed methods for solving the formidable Einstein gravitational field equation on the computer. Many of these computer studies of the Einstein theory have yielded predictions that severely violate our established laws of physics. Such predictions are in strong opposition to the philosophy that Einstein demanded. Einstein insisted that, "A theory, in order to merit confidence, must be based on generalizeable facts".

During Einstein's lifetime, the Einstein general theory of relativity appeared to work, and so Einstein had faith in his theory. If he had lived to experience computer studies of his theory, predicting results that seriously conflict with physical laws, he certainly would have realized that his theory should be revised. When we examine Einstein's scientific philosophy, we can reach no other conclusion.

The Yilmaz Gravitational Theory

This book will demonstrate that the Einstein gravitational field equation has a flaw. This flaw has been corrected by the Yilmaz theory of gravity, which is a refinement of the Einstein theory. Starting in 1958, Prof. Yilmaz has published many technical papers on his theory in prestigious scientific journals. With the Yilmaz theory, the physically impossible predictions that have been derived from the Einstein theory are eliminated.

The gravitational field equation of the Yilmaz theory has an additional term not included in the Einstein theory. Yilmaz has derived a general solution to his gravitational field equation, which makes the Yilmaz theory very much easier to apply than the Einstein theory. A number of alternatives to the Einstein gravitational theory have been proposed by scientists, but the Yilmaz theory is the only relativistic theory of gravity that has a general solution.

The fact that the Yilmaz theory yields this general solution proves that it has profound mathematical integrity. There is great arbitrariness in the application of the Einstein theory, because the Einstein gravitational field equation does not precisely constrain the result. In contrast, the general solution provided by the Yilmaz theory has a unique answer. Since the Yilmaz theory can give only one answer, its predictions should be taken seriously.

Since the 1960's, hundreds of scientists have devoted their careers to the task of solving the formidable Einstein gravitational field equation on the computer. This effort has been directed toward Big Bang cosmology studies, because that is the only area that can use this expertise. If the Yilmaz theory should become widely accepted, this Big Bang research would be obsolete. Therefore, it is not surprising that Big Bang theoreticians oppose the Yilmaz theory.

This book gives simple physical descriptions of the Einstein and Yilmaz theories that can be readily understood by the average reader. This will provide the foundation for unlocking the mysteries of our universe.

The Yilmaz Cosmology Model

In the first paper on his theory, Prof. Yilmaz applied his gravitational theory to a simple cosmology model, which assumes a constant average density of matter throughout the universe that does not change with time. Since then Prof. Yilmaz has ignored this cosmology model and other cosmological applications of his theory, because he came to realize that cosmology can be very speculative. He does not want the speculative aspects of cosmology to obscure the rigorous mathematical character of his gravitational theory. Prof. Yilmaz has concentrated on mathematical analyses to expand his gravitational theory, and on tests that can be performed within our solar system to verify the theory.

Independently of Prof. Yilmaz, the author has extended the original analysis of the Yilmaz cosmology model. As this book will demonstrate, this model yields the following revolutionary picture of our universe:

> The Hubble expansion of the universe is a natural relativistic effect that is directly caused by gravity. To compensate for this expansion, energy that is radiated from stars is converted into diffuse matter in space. This matter gradually coalesces to form new stars and galaxies, and the process continues indefinitely.
>
> The model yields the remarkable prediction that the universe expands about every point, but its over-all size remains constant. This surprising concept is a direct consequence of the relativistic equations of the Yilmaz gravitational theory.
>
> The universe is predicted to be infinitely old, but appears to be eternally young because it is continually changing.

Is this picture of the universe correct? It should be taken seriously for the following reasons. *First,* it is based on a rigorous theory of gravity, unlike other cosmology theories that apply the flawed Einstein gravitational field equation. *Second,* it is consistent with our laws of physics; whereas all variations of the Big Bang theory seriously conflict with physical laws.

Anyone who examines with an open mind the multitude of scientific attempts to explain our universe should realize that none of them make sense. The explanations sound more like science fiction than science. The theme of this book is that cosmological models have failed because essentially all are based on the Einstein general theory of relativity, which does not yield reliable cosmological predictions. The Yilmaz gravitational theory has refined the Einstein theory to correct its weaknesses, and thereby provides the foundation for a common-sense study of our universe.

This book discusses many complex scientific issues, which are presented in a simple manner that can be readily understood by the average reader. The reader should find this book to be intellectually stimulating, but not overpowering.

How Was Our Universe Created ?

xviii How Was Our Universe Created?

Figure 1-1: The M51 Whirlpool galaxy, which is 35 million light years away, resembles our own Milky Way galaxy. The smaller companion galaxy at the top, NGC5195, is partly obscured by dust from a spiral arm of M51.

Chapter 1

Introduction

Our Beautiful and Mystifying Universe

There is probably nothing more beautiful than the stars on a clear moonless night in a rural location that is far from city lights. The most inspiring feature is that pale white pathway across the sky called the Milky Way. Many people today have never seen the Milky Way, because it is obscured by background lights. The Milky Way becomes particularly exciting when we realize what it represents. We are looking at billions and billions of distant stars within our Milky Way galaxy.

Our Milky Way galaxy is similar to the M51 Whirlpool galaxy shown in Fig. 1-1. It has a disk-like shape, with arms spiraling out from a central nucleus. Our sun is a star that is located 2/3 of the distance from the center to the circumference, and lies within a spiral arm. The Milky Way galaxy contains about 100 billion stars. It is so enormous that light takes one hundred thousand years to travel across it.

Prior to the early 1900's it was generally believed that our galaxy comprised the whole universe. As astronomers developed means of measuring stellar distances, they discovered that many of the fuzzy astronomical objects called "nebulae" were actually distant galaxies containing billions of stars like our own Milky Way. They discovered that our Milky Way galaxy is merely one out of billions of galaxies that comprise our universe. The size of our universe is so enormous it seems beyond comprehension.

In 1929, astronomer Edwin Hubble (1889-1953) discovered that our universe is expanding. Except for a few close galaxies, he found that all galaxies are moving away from us at velocities approximately proportional to distance. If we extrapolate galaxy motions backward in time, the whole universe seems to have emerged from a single point with

2 How Was Our Universe Created?

a "**Big Bang**" about 15 billion years ago. This is the foundation for the Big Bang theory.

Although 15 billion years may seem like a long time, it is short when compared to the age of our solar system. There is a great deal of evidence showing that our sun is about 5 billion years old. It seems inconceivable that our whole universe can be only 3 times as old as our sun.

The Einstein Gravitational Theory

In order to develop a quantitative explanation of cosmology, one needs a theory of gravity. The gravitational theory of Isaac Newton gives accurate predictions of the orbits of planets and other bodies within our solar system. However, Albert Einstein showed that there is a very small component in the orbit of the planet Mercury that cannot be explained by Newton's theory, and he predicted this component with his general theory of relativity. The Einstein general theory of relativity is basically a theory of gravity, which approximates Newton's laws in a weak gravitational field.

The gravitational field within our solar system is weak. Although the Newtonian theory gives accurate predictions in a weak gravitational field, it is not adequate to handle the much stronger gravitational fields encountered in a study of our universe. Consequently a theory of cosmology is usually based on the Einstein general theory of relativity.

The Einstein general theory of relativity was a monumental advance in our understanding of physics. This theory involves a large body of very sound scientific principles. However, the calculations of general relativity are very complicated, and this complexity has obscured the fact that Einstein did not quite get the correct answer with his theory.

This theory is specified by the Einstein gravitational field equation, which is a tensor formula that represents a set of 10 independent equations. When the theory is applied to a complicated physical model, the Einstein gravitational field equation can result in millions of terms, and so can only be solved with a computer. Since computers were not available during Einstein's lifetime, Einstein had to restrict the application of his theory to very simple physical models. Consequently Einstein did not realize that his gravitational field equation is not a rigorous solution of the principles of relativity.

Einstein insisted that his theories must agree with observational evidence. He strongly opposed the physically impossible "singularity" concept (derived from his general relativity theory), which predicted that a star with sufficient mass to form a "black hole" must collapse indefinitely until its size shrinks to zero.

Many attempts have been made to use the Einstein theory as a foundation for a theory of cosmology, but they have not resulted in an explanation of our universe that agrees with observational evidence and our laws of physics. A primary reason for this failure is that the Einstein gravitational field equation does not yield reliable predictions in an intense gravitational field.

The Big Bang Explanation of our Universe

Based primarily on computer studies of the Einstein general theory of relativity, tremendous effort has been expended over the past third of a century to explain the Hubble expansion of the universe. These studies have convinced most scientists working on the problem that our universe must have exploded with a "Big Bang" about 15 billion years ago. The analyses seem to lead inescapably to the conclusion that the universe began as an extremely compact mass having an unbelievably high density of matter.

Our universe holds many billions of galaxies, each of which contain luminous material that is equivalent to many tens of billions of stars like our sun. There is probably at least 100 times as much dark matter (which we cannot see) as there is luminous matter. Many respected authorities in astronomy claim that, at the instant of the Big Bang, all of this material in our universe was compressed within an enormously dense mass that was smaller than an inch!

This concept sounds like science-fiction. Nevertheless, we are told that we should accept this bizarre claim with confidence, because it has been endorsed by scientific authorities. We should not be disturbed if it violates our common sense.

As this book demonstrates, there is a great deal of scientific evidence that conflicts with the Big Bang theory. This evidence is being brushed under the rug by those who control the field of astronomy. Open scientific debate is eliminated. Astronomical research, which is funded by enormous amounts of taxpayer money, is rigidly controlled to follow the Big Bang concept. Since the Big Bang is proclaimed to be fact, not theory, any research effort in opposition to the Big Bang is denigrated as being irresponsible. The book presents disturbing examples of this suppression of scientific evidence.

4   *How Was Our Universe Created?*

Need for a Reliable Theory of Gravity

When we examine the Big Bang theory, with its countless variations, and the well-known alternative theories of cosmology, we find that they have a fundamental similarity. Nearly all are based on the Einstein general theory of relativity. However, as this book will demonstrate, the gravitational field equation that specifies the Einstein theory has a flaw. The Einstein gravitational field equation does not yield reliable predictions in an intense gravitational field, and so it cannot provide the foundation for a physically meaningful theory of cosmology.

While performing PhD studies at the Massachusetts Institute of Technology in the early 1950's, Huseyin Yilmaz examined the steps that Einstein took in developing his general theory of relativity. Yilmaz discovered that Einstein had made an unnecessary approximation when calculating the change of wavelength produced by gravity. Yilmaz performed this calculation exactly, and this led directly to a rigorous solution of general relativity. Thus Yilmaz has refined the Einstein theory to develop his theory of gravity. Starting in 1958, Yilmaz has published numerous technical papers on his theory in prestigious scientific journals.

To apply the Einstein theory, one must solve its very complicated gravitational field equation. Yilmaz has derived a general solution to his gravitational field equation, and so the Yilmaz gravitational field equation never needs to be solved. Consequently, the Yilmaz theory is very much easier to apply than the Einstein theory. *The fact that the Yilmaz theory yields this general solution proves that it has profound mathematical integrity.*

Einstein did not like the great complexity of his general theory of relativity. He always tried to achieve as simple a theory as possible. If he had retraced the steps he took to develop general relativity, he probably would have discovered the exact solution that Yilmaz found. This would have given him the simple theory that he wanted, and scientific developments in this field would have been radically different.

There is great arbitrariness in the application of the Einstein theory. Multiple solutions can be derived from the same physical model, depending on what "coordinate" assumptions are made. The Yilmaz theory does not have this weakness. Since it has a definite general solution, it can yield only one answer for a particular physical model. Because of this property, the predictions of the Yilmaz theory are unique, and should be taken seriously. This issue is discussed further in Chapter 9.

It is not surprising that the Yilmaz theory is opposed by the multitude of theoreticians who are using the Einstein theory in Big-Bang computer studies. If the Yilmaz theory should become widely accepted, it would undermine this enormous research effort.

The Yilmaz Cosmology Model

When the Yilmaz theory is applied to cosmology, it yields some remarkable predictions. In the first paper on his theory, published in 1958, Yilmaz applied his theory to a simple cosmology model. The model assumes that the universe has a constant average density of matter that extends to infinity and does not change with time. Yilmaz was surprised to discover that his cosmology model predicts an expanding universe. Relativistic effects due to gravity should make the universe expand, just as Hubble had observed.

Prof. Yilmaz came to realize that cosmology can be very speculative, and so he has ignored cosmology in subsequent papers. He has worked toward mathematical extensions of his theory and experiments that can be performed within our solar system to test the theory. He does not want the speculative aspects of cosmology to obscure the rigorous mathematical character of his relativistic theory.

Operating independently of Prof. Yilmaz, the author is applying the powerful Yilmaz gravitational theory to cosmology. The analysis of the Yilmaz cosmology model was extended. This has shown that relativistic effects distort space in such a manner that the universe expands locally about every point, but over very large distances the universe does not expand. In other words, the Yilmaz cosmology model makes the astounding prediction that our expanding universe does not get any bigger as it expands. **Even though the universe expands everywhere, its overall size remains constant!** We will see that this confusing prediction makes sense when we consider relativistic effects.

Our Picture of the Universe

To investigate our mysterious universe, we must understand the Einstein general theory of relativity and the Yilmaz theory of gravity. Although the mathematical calculations of relativity are complicated, this book demonstrates that relativity principles can be explained in a simple physical manner that is easily comprehended by the average reader.

Armed with this basic understanding of the physical principles that

6   *How Was Our Universe Created?*

underly the Einstein and Yilmaz theories, the reader will have the foundation that is needed to explore the wonders of cosmology. We will examine the predictions derived from the Yilmaz cosmology model and find that it yields the following exciting picture of our universe:

> *The age of our universe is infinite. The universe has always been more or less like we see it today, and will always remain that way. It expands locally, in accordance with the Hubble law, but does not grow any bigger because it does not expand over very large distances. The Hubble expansion is a local relativistic effect that is caused by gravity. Energy radiated from stars is transmitted across the universe to create matter that compensates for the Hubble expansion. This creation of matter forms new stars and galaxies to keep the universe continually changing, so that the universe stays eternally young even though it is infinitely old.*

Is this picture of the universe correct? Since it has evolved from a rigorous theory of gravity, it has a much firmer theoretical foundation than the Big Bang theory or any of its well-known alternatives. It deserves our careful evaluation.

Mankind has continually asked, "How were we created?" and "How was our world created?" The insights that the reader gains from a journey through this book should be a rewarding adventure in the pursuit of these eternal questions.

Chapter 2

The Big Bang Theory

The Big Bang and Black Hole Singularities

We saw in Chapter 1 that the expansion of the universe discovered by Edwin Hubble seems to imply that our universe began as a "point" about 15 billion years ago. This is the basis for the Big Bang theory. Let us now ask the question, "How small was that point?"

Physicists and mathematicians often use the concept of a "singularity" in their analyses, which is a point having zero size, and can represent an infinite density of matter. When normally used, a "singularity" is a mathematical abstraction that approximates the real world. However, when Big Bang theorists use the term "singularity" in describing cosmology, they are referring to physical reality. They literally mean a physical condition where the density of matter is infinite or practically infinite.

Stephen Hawking has been widely proclaimed to be one of the world's leading experts on cosmology. His concepts were described in a 1997 NOVA television documentary, called "Stephen Hawking's Universe" presented by the Public Broadcasting System (PBS). A related book with the same name was written by Filkin [7]. This book states the following on page 104:

> "Stephen (Hawking) and Roger Penrose published a paper in 1970 which proved that, if Einstein's mathematics were correct, a singularity had to result from a black hole, and had to exist at the start of the universe. ···· The paper argued that if relativity as explained by Einstein is correct — and all of the evidence from observation seems to keep confirming it — then the universe must have started with a Big Bang explosion out of a singularity. The equations do not allow an alternative."

8 How Was Our Universe Created?

The *Scientific American* issue of January 2001 was devoted primarily to the "Brave New Cosmos", which describes various aspects of the Big Bang theory. The featured expert on cosmology was P. James E. Peebles, who summarized and evaluated the different versions of the Big Bang. The October, 1994 issue of *Scientific American* presented an article on the Big Bang theory written by Peebles and others [8], which began with:

"At a particular instant roughly 15 billion years ago, all of the matter and energy we can observe, concentrated in a region smaller than a dime, began to expand and cool at an incredibly rapid rate."

It seems incredible to claim, as this does, that our whole universe, containing many billions of galaxies, was once compressed within a "region smaller than a dime". Nevertheless most of our public has accepted the pronouncements of Big Bang theorists, which grossly conflict with our laws of physics and our common sense.

The Big Bang and Black Hole concepts are closely related, because both involve a physical "singularity". The basic principle of the black hole is well known to the general public. There have been many science-fiction discussions of the imaginary experience of "falling into a black hole". We are told that when a star reaches a critical density of matter, it becomes a black hole that is surrounded by a spherical "event horizon" surface, over which the speed of light is zero. The force of gravity is so tremendous that neither light nor matter can escape from inside this event-horizon sphere.

Although the basic concept of the black hole is well known, few understand that the star inside a black-hole event-horizon must collapse until its size shrinks to zero. Let us examine the basis for this prediction.

The black hole concept evolved from the Einstein general theory of relativity. The equations of general relativity, which were presented by Einstein in 1916, are extremely complicated, and Einstein could only obtain approximate solutions of his theory. The brilliant physicist, Karl Schwartzschild, derived an exact solution, which characterized the gravitational effects of a star. This Schwartzschild solution was also published by Einstein in 1916. Unfortunately Karl Schwartzschild died suddenly from disease even before his famous solution was printed. He had been a German officer on the Russian front during World War I.

The Schwartzschild solution to the Einstein theory reaches a mathematical limit when the ratio of mass to radius of a star exceeds a

2. The Big Bang Theory 9

critical value that is 240,000 times the ratio for our sun. When this Schwartzschild limit is exceeded, the pressure inside the star should become "imaginary", which is an impossible condition. This indicates that the Schwartzschild analysis does not yield a solution when this mathematical limit is exceeded. Einstein ignored the Schwartzschild limit, because he was interested in applying his theory to physically meaningful situations like our solar system.

In 1939, J. R. Oppenheimer (who later directed the atomic bomb project), wrote a paper on this problem in the *Physical Review*, in cooperation with H. Snyder [9]. The Schwartzschild analysis assumed a "static" condition, meaning that the size of the star is constant. Oppenheimer and Snyder found that the Einstein theory can yield a real value for pressure inside the star, when the critical Schwartzschild mass-to-radius limit is exceeded, provided that one assumes a "dynamic" condition, in which the size of the star continually decreases. Oppenheimer and Snyder concluded that when the ratio of mass to radius of a star exceeds the critical Schwartzschild limit, the star must shrink indefinitely.

This indicates that, if the Schwartzschild limit is exceeded, a star should become a black hole, surrounded by an event horizon. The star inside this event horizon should shrink until its size becomes zero. Since its mass does not change, its density of matter should become infinite.

This 1939 article by J. R. Oppenheimer and H. Snyder [9] presented an analysis based on the Einstein general theory of relativity, which concluded with:

"When all thermonuclear sources of energy are exhausted, a sufficiently heavy star will collapse. Unless fission due to rotation, the radiation of mass, or the blowing off of mass by radiation, reduce the star's mass to the order of that of the sun, this contraction will continue indefinitely."

This statement claimed that when the mass-to-radius ratio of a star exceeds the Schwartzschild limit, the star must contract "indefinitely" until its size shrinks to zero. The mass of the star does not change, and so the density of matter becomes infinite, and the star forms a singularity. Albert Einstein strongly opposed this prediction, derived from his theory, because it is physically impossible for a star to be squeezed into a singularity. A month after the Oppenheimer-Snyder article was published, Einstein responded with an extensive analysis in *Annals of Mathematics* [10], but the Einstein article politely did not

10   How Was Our Universe Created?

specifically refer to the Oppenheimer-Snyder article. Einstein concluded with:

> *"The essential result of this investigation is a clear understanding as to why the 'Schwartzschild singularities' do not exist in physical reality. Although the theory here treats clusters whose particles move along circular paths, it does not seem to be subject to reasonable doubt that more general cases will have analogous results. The 'Schwartzschild singularity' does not appear for the reason that matter cannot be concentrated arbitrarily. And this is due to the fact that otherwise the constituting particles would reach the velocity of light."*

A few years later Oppenheimer became manager of the Manhattan atomic bomb project and never pursued this issue again. Neither did any other scientist while Einstein was alive.

The physically impossible concept of a star collapsing into a singularity having infinite density of matter was drastically inconsistent with Einstein's philosophy. The thinking of Albert Einstein on such issues is explained in a recent and very thorough biography of Einstein, originally written in German by Folsing [11], who states (p. 381):

> *"Some of Einstein's admirers were tempted to see the general theory of relativity as a triumph of speculation over empiricism. This kind of misunderstanding made Einstein 'downright angry' [who said] 'This development teaches us something entirely different, indeed almost the opposite, namely that a theory, in order to merit confidence, must be based on generalizeable facts'. . . To Einstein, facts were not only the starting point of his theory but also the keynote of any test of it."*

Like the black hole, the Big Bang theory also involves a singularity, in which the density of matter is practically infinite. We are fed authoritative statements by learned scientists asserting the validity of the Big Bang theory, along with predictions that radically violate our established laws of physics.

Big Bang theorists use the name of Albert Einstein to support their position, because their analyses are based on the Einstein theory. They fail to admit that their "singularity" concepts radically conflict with Einstein's philosophy. You should use your own intelligence to evaluate this issue, and not be intimidated by the scientific establishment. When you take a skeptical attitude toward the Big Bang theory, you are in good company. You are agreeing with Albert Einstein.

Measuring the Hubble Expansion of the Universe

As stated in Chapter 1, astronomer Edwin Hubble discovered in 1929 that our universe is expanding. Galaxies are moving away from us at velocities approximately proportional to distance. To understand this finding, let us examine the methods that Hubble used to measure the distances and velocities of galaxies.

The Hubble constant, denoted H_0, is the ratio of galaxy velocity to distance. Measurements of the Hubble constant have improved greatly in recent years, assisted in particular by data from the Hubble space telescope. The average value derived from recent data is about 20 km/sec (kilometers per second) per million light years. Thus a galaxy that is 10 million light years away recedes from us at a velocity of about 200 km/sec. The first data published by Hubble yielded an expansion rate that was 8.5 times greater than this, because of errors in his estimates of galaxy distance. Let us consider the steps involved in determining the Hubble constant.

The radial velocity of a star means its velocity in the radial direction, which is the component of velocity either toward us or away from us. The radial velocity of a star can be measured accurately from its spectrum. The velocity of a star in the tangential direction (perpendicular to the radius) is very difficult to measure.

An excited atom near the surface of a star radiates a unique pattern of spectral lines, which fall at known wavelengths for a given element. The primary lines in a star spectrum are usually due to hydrogen and helium, but spectral lines for other elements are commonly found.

When the star moves in a radial direction, these spectral lines are shifted by an amount approximately proportional to the radial velocity of the star. This shift was first predicted by Christian Doppler in 1842, and so is called the Doppler effect. If the star is moving toward us, the spectral lines are shifted toward the blue end of the spectrum (toward shorter wavelengths). If the star is moving away from us, the lines are shifted toward the red end of the spectrum (toward longer wavelengths). Hence, one can determine the radial velocity of a star by observing the wavelength shift of its spectral lines. The ratio of the wavelength shift divided by the normal wavelength is approximately equal to V/c, where V is the radial velocity and c is the speed of light.

By examining the spectra emitted by galaxies, Hubble found that,

12 How Was Our Universe Created?

except for a few close galaxies, the spectra are all shifted toward the red, which indicates that the galaxies are all moving away from us. A wavelength shift toward the red is called a "redshift". Hubble developed means of estimating the distances of galaxies, and discovered that the redshift of a galaxy is approximately proportional to its distance. This finding is generally interpreted to mean that our universe is expanding at a constant rate. Galaxies appear to be moving away from one another at velocities proportional to the distance between them.

The concept of an expanding universe can be illustrated by assuming that a rubber band, with perfect elasticity, is under tension and is being stretched at a constant rate. At an instant of time (called t_1) mark a series of dots on the band separated from one another by 10 mm, and label these dots as follows:

```
D'   C'   B'   A    B    C    D    E
*    *    *    *    *    *    *    *
```

The distance from A to B is 10 mm, from A to C is 20 mm, from A to D is 30 mm, etc. Now look at the band later (at time t_2) when it has stretched by 10 percent, so that the distance between neighboring dots has increased to 11 mm. The distance from A to B is now 11 mm; from A to C is 22 mm; from A to D is 33 mm, etc. Between time t_1 and time t_2, point B moves 1 mm away from A, point C moves 2 mm away from A, point D moves 3 mm away from A, etc. The relative velocity between points along the band is proportional to the distance between the points. This indicates that the rubber band is being stretched at a constant rate.

We can extend this analogy to two dimensions by considering a perfectly elastic balloon that is being blown up at a constant rate. Mark on the surface of the balloon an array of dots separated by equal distances. As the balloon expands, the dots recede from one another at velocities proportional to the distances between the dots.

A key aspect of Hubble's discovery was his development of techniques for determining galaxy distances. Distance measurement in astronomy is a complicated process, and is subject to considerable error. Let us consider the steps involved in this measurement of distance.

The first step is to use parallax to measure the distances of nearby stars. Parallax is the principle that our eyes and brain use to achieve depth perception. The images that our two eyes receive are not exactly the same. The brain compares the two images, and thereby can distinguish close objects from those that are further away.

As the earth rotates around the sun, the parallax effect causes the images of nearby stars to shift relative to the distant stars. The parallax shift is inversely proportional to distance, and is equivalent to one arc second for a distance of 3.26 light years. If the star image shifts by 0.1 arc second over the year, we know that the star is located 32.6 light years away. By this means, astronomers can measure stellar distances out to about 100 light years with ground-based telescopes. The Hubble space telescope can make much more accurate parallax measurements, because it does not experience the blurring effect of the atmosphere. With the Hubble telescope, stellar distances have been measured more accurately from parallax, and out to greater distances.

The stars that could be measured by the parallax method were placed in different categories based on their spectra. For certain spectral types, the power radiated by the star is approximately constant, and so this type of star can be used as a "standard candle". If one finds a more distant star of the same spectral type, one can estimate its distance by recognizing that the light power that is received from a star is inversely proportional to the square of its distance.

An accurate standard candle for astronomy was developed by astronomer Henrietta Leavitt (1868-1921) in 1908, based on Cepheid variable stars. These are stars that vary periodically in luminosity, and are named after the star Delta Cephei. Leavitt observed the Cepheid variable stars located in the Small Magellanic Cloud, which is a cluster of stars about 300,000 light years from the earth. The Small Magellanic Cloud can be seen only in southern latitudes.

All of these stars in the Small Magellanic Cloud are at approximately the same distance, and so can be compared directly with one another. Leavitt found that the period of oscillation of a Cepheid variable star is directly related to the light power that it radiates. By examining nearby Cepheid variable stars, the distance to which can be measured by parallax, she was able to relate the period of oscillation of a Cepheid variable star to the absolute light power that the star radiates. This measurement technique developed by Henrietta Leavitt was the key that allowed Edwin Hubble to make his revolutionary discovery. Hubble had the advantage that he was using the new telescope at Mount Wilson Observatory, which had much greater resolution than earlier telescopes.

Hubble examined Cepheid variables in the galaxy M31 of the constellation Andromeda, which is 2.3 million light years away, and in the M33 galaxy of the constellation Triangulum, which is 2.6 million light years away. In this way he estimated the distances to M31 and M33. He examined the brightest stars in these galaxies, which he believed to be super-giant stars. He assumed that the brightest stars in

14 How Was Our Universe Created?

more distant galaxies would radiate the same power as those in the M31 and M33 galaxies. By this means, he estimated distances to several galaxies more distant than M31 and M33, which were too far away to observe Cepheid variables.

Hubble categorized these galaxies into different types depending on their shapes. He assumed that all galaxies of a given shape have approximately the same dimensions. With this assumption, he estimated the distances to galaxies that were much further away. He derived his universe expansion rate by comparing the redshifts with the estimated distances of these very distant galaxies.

The expansion rate that Hubble measured was a factor of 8.5 times greater than the value generally accepted today. One serious problem was that there are two types of variable stars: the Cepheid variable stars and the RR Lyrae stars. This fact was not discovered until the 1950's. This effect made an error of 2 in Hubble's estimates of the distances to the M33 and M31 galaxies. There were many other sources of error that contributed to an inaccuracy factor of 8.5 in the Hubble constant.

Recent measurements of the Hubble constant have been based primarily on Cepheid variable stars, and supernovas. By means of the Hubble space telescope, Cepheid variable stars can be observed in much more distant galaxies. A supernova is an exploding star that briefly displays a brightness that is millions of times greater than a normal star. There is a particular type of supernova, called *Type 1a*, which has a predictable brightness, and can be distinguished by its spectrum. Type 1a supernovas can be observed in very distant galaxies. The absolute brightness of this supernova type has been calibrated from nearby galaxies by examining Cepheid variables in the same galaxy. However, this comparison has not been made accurately, and so there is still appreciable uncertainty in the estimates of the Hubble constant.

Astronomers usually express the Hubble constant in terms of the parsec, which represents 3.26 light years. One parsec is the theoretical distance of a star that would exhibit an annual parallax shift of one arc second. A Hubble constant of 20 km/sec per million light years can be expressed as 65 km/sec per megaparsec, where the term "megaparsec" means "million parsecs".

This book assumes a Hubble constant of 25 km/sec per million light years, rather than 20 km/sec per million light years. This was the average value that was generally assumed a few years ago. Since there is still uncertainty in the Hubble constant, the author prefers to continue using this value. The conclusions of this book are not significantly affected by the precise value of the Hubble constant.

Development of the Big Bang Theory

The Apparent Age of the Universe

We have seen that the astronomical observations by Edwin Hubble, when corrected by recent data, appear to show that the universe is expanding at a rate of 20 km/sec per million light years. This conclusion is based on observations of the redshift of light received from galaxies. Although this redshift is generally interpreted to be a Doppler effect caused by a receding galaxy velocity, other effects can also cause a redshift. Consequently we cannot be absolutely sure that the Hubble redshift actually represents an expansion of the universe. Alternate interpretations of the Hubble redshift will be considered later. Nevertheless, for the moment we assume that the Hubble expansion of the universe is real.

Since the speed of light c is 300,000 km/sec, a Hubble constant of 20 km/sec per million light years can be expressed as c/15,000 per million light years, or c/15 per billion light years. This is equivalent to an expansion rate of the speed of light c over a distance of 15 billion light years. If we extrapolate this expansion backward in time, the universe seems to have begun as a single point 15 billion years ago.

This book calls 15 billion years the apparent age of the universe, which is denoted T_0. If we assume that the whole universe expands exactly in accordance with a Hubble constant of 20 km/sec per million light years, and has always done so, the universe would have begun as a point, 15 billion years ago. This book considers the age of the universe to be infinite, and that the apparent age T_0 is merely a convenient parameter to characterize the Hubble expansion.

In its simplest form, the Big Bang theory would regard T_0 to be the true age of the universe since the Big Bang, and so a Hubble expansion rate of 20 km/sec per million light years would imply a universe age T_0 of 15 billion years. However, to explain the Big Bang theory, theorists have come to believe that the expansion of the universe was not exactly constant. Consequently, the true ages of the universe that have been computed from various versions of the Big Bang theory are not exactly equal to the apparent age T_0.

Historical Development of the Big Bang Theory

In 1917, one year after Albert Einstein presented his general theory of relativity, he published a mathematical model of the universe that was

16 How Was Our Universe Created?

based on his theory. William de Sitter (1872-1934), a Dutch astronomer, published a similar model in that same year. These models assumed a "static" universe that did not vary with time. Einstein had included a "cosmological" term in the equation for his theory, so that his theory could allow a static cosmological solution.

When Hubble discovered in 1929 that the universe is expanding, Einstein realized that a static solution of his theory is not consistent with an expanding universe. A dynamic solution is required, and a dynamic solution would not need his "cosmological term". At that time Albert Einstein disavowed his cosmological term, claiming it to be, "one of the greatest mistakes of my life".

Time-varying (or "dynamic") cosmology models based on the Einstein theory were first developed independently by Alexander Friedmann (1888-1925) in 1922 and by Georges Lemaitre (1894-1966) in 1927. Friedman was a Russian meteorologist and mathematician, and Lemaitre was a Belgian cleric of the Roman Catholic Church. Friedman died in 1925 after being chilled in a weather balloon.

The dynamic cosmology model by Lemaitre assumed that the universe began as a highly compact body. During the 1930's Lemaitre was the primary spokesman for this concept. A problem at that time was that the accepted Hubble constant implied an apparent universe age of only 2 billion years.

During World War II, little attention was paid to this matter. After the war, the chief proponent of the Lemaitre concept was George Gamow (1904-1968), a physicist who had worked on the Manhattan project that developed the atomic bomb. He wrote several technical papers on cosmology, and books for the general public that became very popular. By 1950 it was recognized that the original Hubble expansion rate was too high. The new Hubble constant implied a universe age of about 5 billion years.

At that time the telescope at Mt. Wilson Observatory could observe galaxies to a distance of about 500 million light years. Gamow estimated that all of the matter observable over this 500 million light-year distance was originally compressed within a volume having 8 times the diameter of our sun.

Gamow theorized that the elements that we observe today were created in this embryonic universe. With its highly dense state, he believed that this young universe could have exhibited the extreme temperatures and pressures required to create with nuclear reactions the various elements that are found on earth. However, later studies have concluded that most of our elements were probably generated within exploding stars called supernovas.

An alternate cosmological theory, called the "Steady-State Universe theory", was first proposed in 1946 by the British astronomer Fred Hoyle (1915-). A modified version of this concept was developed by Hermann Bondi and Thomas Gold. This theory assumes that the age of the universe is infinite. As the universe expands, matter is created to compensate for the universe expansion. This theory is explained in more detail in Chapter 3.

In a criticism of Gamow's theory of cosmology, Fred Hoyle dubbed the initial explosion of the universe as the "Big Bang". This name stuck, and since that time the Lemaitre-Gamow cosmology model has been officially known as the "Big Bang theory".

Up through the 1950's, studies of the Big Bang theory were limited. An enormous increase of Big Bang studies occurred with the development of powerful computers, which became widely available in the 1960's.

Effect of the Computer on Cosmological Studies

Albert Einstein died in 1955, which was before computers became generally available. Unless the physical model is very simple, the equations of general relativity can result in millions of terms, and so cannot be solved analytically. Consequently, Einstein had to limit his general relativity studies to simple cases.

When powerful computers became widely available in the mid 1960's, many physicists, mathematicians, and engineers in academic positions began to perform computer studies of general relativity. With computers, the formidable equations of general relativity could now be solved in a manner unheard of in Einstein's day.

There was strong economic incentive to perform these studies. Universities are all searching for prestige, which is gauged primarily by publications. The motto among university professors is "Publish or Perish". Since the theory of Albert Einstein is treated with awe, anyone who can obtain a publication based on the Einstein general theory of relativity is sure to achieve strong commendation from his superiors.

The basic theory of relativity that Einstein presented in 1905, which is called *Special Relativity*, has wide applicability. However, the much more complicated *General Theory of Relativity* that Einstein published in 1916 has very limited practical application. Einstein developed his general theory in order to provide a solid theoretical foundation for special relativity. Special relativity does not apply exactly under

18 *How Was Our Universe Created?*

conditions of acceleration or in a gravitational field, but it yields a very accurate approximation in nearly all experiments encountered within our solar system. In those rare cases within our solar system where general relativity is needed, a simple physical model can be used, because the gravitational field within our solar system is weak.

For these reasons, there is very little need for computer studies of general relativity that are related to experiments performed within our solar system. The only meaningful area where these computer studies can be applied is cosmology. Therefore, starting in the mid 1960's there was an enormous increase in theoretical Big Bang cosmology research, based on computer analyses of the Einstein general theory of relativity.

Eric Lerner wrote an excellent book refuting the Big Bang theory, titled *The Big Bang Never Happened* [4], which was published in 1991. He was assisted greatly in this effort by Hannes Alfven (1908-1995), who received a Nobel prize for his pioneering work in plasma physics. This book will be discussed in more detail later. This tremendous increase of theoretical Big Bang research is summarized as follows:

As Lerner [4] (p. 153, 154) explains, economic forces caused an enormous expansion of theoretical cosmology research. The annual number of cosmology papers skyrocketed from 60 in 1965 to 500 in 1980, and the increase was almost entirely in purely theoretical work. In 1980, 95 percent of these papers were devoted to various mathematical models, such as the "Bianchi type XI universe".

In Lerner's words: during the 1970's, "the field of cosmology was transformed from a small group of squabbling theorists trying to develop theories that would match observations, to a huge phalanx of hundreds of researchers, virtually all united in their basic assumptions, and preoccupied with the mathematical nuances of the underlying theory." This enormous theoretical effort in cosmology has continued to this day, and international conferences on cosmology are now held about once a month to discuss the latest "hot" Big Bang model.

We have seen that Einstein strongly opposed the Schwartzschild singularity concept associated with the black hole, because he realized that a singularity strongly violates our established laws of physics. With theoretical analysis, Einstein convinced himself that his theory does not predict a black hole singularity.

Einstein's claim could not be disputed at that time, because analytical studies of the Einstein theory are very difficult. When computers were applied to this problem in the 1960's (after Albert Einstein had died), the studies yielded strong evidence that Einstein was wrong. In order to satisfy the equations of the Einstein theory, the computer studies apparently proved that a star must collapse indefinitely until it becomes a singularity having infinite density of matter, if its mass-to-radius ratio exceeds the critical Schwartzschild limit.

This is certainly an important finding, but what do we conclude from it? If Albert Einstein had lived to experience computer studies of his theory, it seems inconceivable that he would have accepted the physically impossible conclusion that a black-hole star actually shrinks into a singularity having zero size and infinite density of matter. The philosophy that Einstein maintained throughout his life was radically inconsistent with this non-physical conclusion. Such computer evidence would certainly have convinced Einstein that there must be something wrong with his theory.

With the wild Big-Bang predictions calculated from the Einstein theory, Big Bang proponents claim that they are following in the footsteps of Einstein. To be sure, they are using the Einstein theory, but they are grossly violating the philosophy that Einstein demanded throughout his life. As shown earlier in this chapter, in his extensive biography on Einstein, Folsing [11] quoted the following words of Einstein, "This development teaches us something entirely different, indeed almost the opposite, namely that a theory, in order to merit confidence, must be based on generalizeable facts". In Folsing's words, "To Einstein, facts were not only the starting point of his theory but also the keynote of any test of it."

The extensive computer studies of the Einstein theory were applied to cosmology and led inescapably to the Big Bang theory, with its associated singularities. Let us reexamine the following statement given earlier in this chapter, obtained from Filkin's book, *Stephen Hawking's Universe* [7]:

"Stephen (Hawking) and Roger Penrose published a paper in 1970 which proved that, if Einstein's mathematics were correct, a singularity had to result from a black hole, and had to exist at the start of the universe."

Gamow had estimated that a portion of our universe that is one billion light years in diameter should have begun as a body 8 times the diameter of the sun. With the application of the singularity concept, many Big

Bang cosmologists now insist that our present observable universe, 30 billion light years in diameter, was squeezed into a space smaller than a dime.

Cosmic Microwave Background Radiation

An important milestone in the development of the Big Bang theory occurred in 1964, when Arno Penzias and Robert Wilson, two physicists working at Bell Laboratories, discovered cosmic microwave background radiation. These physicists were performing measurements on a sensitive microwave antenna that had been developed for use in satellite communication. When communication satellites generated greater power, there was no longer a communication need for this sensitive instrument, and so the antenna was redirected toward basic research. Penzias and Wilson discovered spurious signals in their antenna at microwave frequencies, which they could not explain. These signals were due to cosmic radiation coming from all directions in space. Big Bang theorists had predicted that such radiation should have been generated by the Big Bang.

Gamow had predicted that optical radiation from the early universe should still be observable today. With the expansion of the universe, this cosmic optical radiation should be reduced in frequency, and should be observable today at microwave frequencies. He predicted cosmic radiation coming from all directions that is equivalent to the radiation from an ideal blackbody. Gamow made several estimates of the effective temperature of this cosmic blackbody radiation, which varied from 5 degrees Kelvin to 20 degrees Kelvin.

Kelvin temperature is measured above absolute zero temperature, which is approximately -273 degrees Celsius. A body at a temperature of 3 degrees Kelvin is 3 degrees Celsius above absolute zero, and so is at a temperature of -270 degrees Celsius. One should not think of how very "cold" a body of this temperature may be in space, because temperature in the vacuum of space has a different physical meaning than here on earth. Temperature in space basically describes the random velocities of the diffuse gas molecules. At absolute zero temperature, the random velocity of gas molecules is zero.

A body at room temperature continually radiates energy into its environment in terms of heat. It also absorbs heat from the environment. Heat radiation is like light, except that has a longer wavelength. The blacker the surface of an object, the better it absorbs radiation, and the better it radiates energy into the environment. A blackbody is a physical

2. The Big Bang Theory 21

idealization that radiates the maximum possible energy from a body at a particular temperature. If one knows the temperature of an ideal blackbody, one knows the spectrum and intensity of the radiation. Intensity is the power that is radiated from a body per unit of surface area.

One can build a nearly ideal blackbody by machining a spherical cavity inside a block of metal, and cutting a small hole into that cavity. Inside the cavity, radiation is emitted from each portion of the spherical surface, and continually reflects off other surfaces, until a small amount escapes out of the hole. The energy escaping from the hole is close to ideal blackbody radiation. The intensity of the blackbody radiation from the hole is always greater than the intensity radiated from the outer surface of the metal block. A blackbody is an idealized physical concept, yet it can be closely approximated by physical equipment.

Intensity of radiation varies as the fourth power of temperature, and so the 4-to-one temperature range from 5 to 20 degrees Kelvin predicted by Gamow corresponds to an intensity ratio of 4^4, which is 256. Thus the Gamow prediction of blackbody radiation was not very precise. There was an uncertainty factor of 256 in the predicted intensity of the blackbody radiation.

In the early 1960's, Prof. Robert Dicke of Princeton University was studying the Big Bang theory. One of his graduate students, P. James E. Peebles, investigated Gamow's prediction of cosmic microwave background radiation, and considered building an antenna to measure this radiation. Peebles estimated that the blackbody temperature should be 30 degrees Kelvin. Then he discovered that Penzias and Wilson had already measured similar radiation in their antenna. The antenna had detected unexplained electrical disturbance signals that peaked at a frequency corresponding to a blackbody temperature of 3.5 degrees Kelvin.

This discovery of the cosmic microwave background radiation was loudly publicized by Big Bang theorists. Since this cosmic radiation had been predicted by the Big Bang theory, it was proclaimed to be proof that the Big Bang theory was correct. However, the Big Bang proponents fail to mention that this radiation was only predicted in a qualitative sense. Since radiation intensity varies as the fourth power of temperature, the 30 degree Kelvin estimate by Peebles corresponds to a radiation intensity that is 5000 times greater than the intensity for a blackbody at 3.5 degrees Kelvin.

In 1948, Ralph Alpher and Robert Herman, two graduate students working with George Gamow, had published a paper in the scientific journal *Nature* that predicted cosmic microwave radiation corresponding to a blackbody temperature of 5 degrees Kelvin. Since this is the closest estimate to the measured temperature, it is the value generally quoted by Big Bang proponents.

Much more accurate measurements of this cosmic radiation were obtained from the Cosmic Background Explorer (COBE) satellite in 1989. This satellite measured cosmic microwave radiation coming uniformly from all directions that corresponded very accurately, in intensity as well as in spectrum, to the radiation from an ideal blackbody at a temperature of 2.73 degrees Kelvin.

The Big Bang theory claims that this cosmic blackbody radiation is the cooled relic of optical radiation that was emitted from the early universe about 300,000 years after the Big Bang. However, an important weakness of this assumption is that the cosmic radiation emanates with extreme uniformity from all directions. Estimates made in 1970 had predicted that energy variations of one part in 1000 in the cosmic radiation would be needed for matter in the universe to have condensed into any structure at all. In the data obtained from the COBE satellite, the energy of radiation received from different directions varies by only a few parts in 100,000.

This implies extreme uniformity of the universe at this early period. How did this highly uniform early universe create what we observe today? Not only is our universe separated into galaxies, but the galaxies are not spaced uniformly. As we will see, the galaxies are arranged into long curling filaments, which form huge ribbon-like structures that are typically 100 million light years thick and one billion light years long.

Cosmic microwave radiation is claimed to have caused the strong support of the Big Bang theory that occurred after 1964. However, this is doubtful. The primary cause for that support was certainly economic. Many hundreds of scientists found that computer studies of the Einstein general theory of relativity represent a rewarding area of research to advance their careers. The Big Bang theory was the only problem to which these computer studies could be applied. Therefore momentum behind the Big Bang theory exploded with a loud bang.

Evidence against the Big Bang

Many scientists with strong credentials are convinced that the Big Bang theory gives an explanation of the universe that best agrees with scientific evidence. However, this does not justify treating the Big Bang theory as fact. This monolithic approach has eliminated scientific debate

2. The Big Bang Theory   23

in the field of astronomy and so is destroying astronomical research.

The Editorial by Professor Geoffrey Burbidge

The case against the Big Bang was expressed eloquently by Professor Geoffrey Burbidge in an editorial article of the February 1992 *Scientific American* magazine [12]. Professor Burbidge is the former director of the Kitt Peak National Observatory, and is presently Professor of Astrophysics at the University of California in San Diego. Prof. Burbidge began his editorial with:

> *"Big bang cosmology is probably as widely believed as has been any theory of the universe in the history of Western civilization. It rests, however, on many untested, and in some cases untestable, assumptions. Indeed, big bang cosmology has become a bandwagon of thought that reflects faith as much as objective truth."*

He went on to say

> *"Younger cosmologists are even more intolerant of departures from the big bang faith than their more senior colleagues are. Worst of all, astronomical textbooks no longer treat cosmology as an open subject. Instead the authors take the attitude that the correct theory has been found."*

Burbidge then explained the basic reasons for the bandwagon mentality:

> *"Powerful mechanisms encourage this conformity. Scientific advances depend on the availability of funding, equipment, and journals in which to publish. Access to these resources is granted through a peer review process. Those of us who have been around long know that peer review and the refereeing of papers have become a form of censorship. It is extraordinarily difficult to get financial support or viewing time on a telescope unless one writes a proposal that follows the party line."*

> *"A few years back, Halton C. Arp was denied telescope time at Mount Wilson and Palomar Observatories because his observing program had found, and continued to find, evidence contrary to standard cosmology."*

"Unorthodox papers often are denied publication for years or are blocked by referees. The same attitude applies to academic positions. I would wager that no young researcher would be willing to jeopardize his or her scientific career by writing an essay such as this."

The case of Halton Arp will be discussed later. Burbidge concluded with the following:

"The big bang ultimately reflects some cosmologist's search for a creation and a beginning. This search properly lies in the realm of metaphysics, not science."

The Big Bang Age Dilemma

Eric Lerner wrote a popular book published in 1991, titled *The Big Bang Never Happened* [4]. He was strongly supported in this book by Nobel laureate Hannes Alfven (1908-1995), who was the father of modern plasma physics. The stimulus for the book was the strong opposition that Hannes Alfven and other plasma physicists had received in their attempts to publish papers in astronomical journals that related plasma physics to cosmology. As we will see, there is strong evidence that plasma physics effects have strongly influenced the development of our solar system, and have caused the rotation of galaxies and stars. Nevertheless, these papers were continually rejected by the Big-Bang establishment, which controls astronomical literature and research.

Lerner [4] gives a detailed criticism of Big Bang research, which is only briefly summarized here. His book is highly recommended to help the reader understand what is happening to astronomy.

Probably the most serious weakness of the Big Bang theory is the universe age dilemma. According to the Big Bang theory, the age of the whole universe can be no greater than about 15 billion years. This is only 3 times the age of our sun, which is 5 billion years old. There is evidence that some stars in our galaxy are nearly 15 billion years old. At one time it appeared that some stars are older than our universe. More recent measurements have reduced the Hubble constant so that this embarrassing condition no longer exists. Nevertheless, there is little margin in the calculations. The development of stars and galaxies must have proceeded in an extremely efficient manner in order for our whole universe to have been formed within 15 billion years.

The Big-Bang age dilemma is actually much worse than this

discussion suggests. As explained by Eric Lerner [4] (pps 15-32), recent astronomical studies have shown that the universe is not at all uniform. In 1986, Brent Tulley, a University of Hawaii astronomer, (assisted by J. R. Fischer) found that almost all galaxies within 1.5 billion light years are concentrated into huge ribbons, about one billion light years long, 300 million light years wide, and 100 million light years thick. These ribbons are made up of curling filaments, a few million light years thick, which extend for hundreds of millions of light years

This study evolved from a mapping of individual galaxies out to 100 million light years. In making this map, it was assumed that the redshift of a distant galaxy specifies its distance. From this map Tulley and Fischer found (with about 20 exceptions) that all of the thousands of galaxies are concentrated into filaments, a few million light years across. These filaments extend for hundreds of millions of light years, beyond the limits of the map. While performing a later mapping study, astronomer Margaret Haynes concluded after examining the curling galactic filaments, *"The universe is just a bowl of spaghetti."*

To expand his study to 1.5 billion light years, Tulley mapped clusters of galaxies, because there are millions of individual galaxies within that range, too numerous to be mapped individually. From this cluster map Tulley discovered his huge ribbons.

About 1990, Tulley's findings were confirmed by several astronomer teams. The most dramatic is by Margaret J. Geller and John P. Huchra of the Harvard Smithsonian Center for Astrophysics, who are mapping individual galaxies out to 600 million light years, about 200 times as many as in the Tulley-Fischer map of individual galaxies. In their preliminary results, they displayed what they call the "Great Wall", a huge ribbon of galaxies stretching across the region mapped, a distance of 700 million light years. This ribbon, which is 200 million light years wide and 20 million light years thick, corresponds closely to a ribbon mapped by Tulley using clusters of galaxies. Galaxy density inside the ribbon is 25 times greater than outside.

These results seriously contradict the Big Bang theory in two ways: (1) it would probably take at least 150 billion years to form these gigantic structures; and (2) the Big Bang theory predicts a very uniform universe, not the spaghetti-like and ribbon-like structures that are observed.

Lerner [4] (page 23) explains the universe age problem. Except for the general Hubble expansion of the universe, the maximum local velocity of all galaxies is less than 1000 km/sec, which is 1/300 of the speed of light. Since the time of the postulated Big Bang, a galaxy could move only a distance of 15 billion light years divided by 300, which is 50 million light years. However, if the universe were uniform after the Big Bang, galaxies would have had to move 270 million light years to form the huge ribbons. The age discrepancy is worse than these numbers suggest, because time must be allowed for a galaxy to accelerate and decelerate.

Mythological Philosophy of Big Bang Research

The cosmic microwave background radiation, which is used as primary evidence to support the Big Bang theory, is actually a serious liability to the theory, because it is far too smooth and uniform. It follows the theoretical blackbody spectrum to very high accuracy, and the radiation arrives with high uniformity from all directions. If this microwave radiation is the cooled relic of light radiated from the very hot matter existing 300,000 years after the Big Bang, the universe must have been extremely uniform at the time the energy was radiated. How did the universe change from that extremely uniform state to the spaghetti-like and ribbon-like universe of today, during the short time following the Big Bang? The 15 billion year age of the universe claimed by Big Bang proponents is at least a factor of 10 too short.

Let us ask the more realistic question: What process could cause galaxies to arrange themselves into these spaghetti-like and ribbon-like structures? To this question, Lerner [4] (pps. 39-49) gives a clear scientific answer. For many years, Hannes Alfven, a Swedish Nobel laureate and virtual founder of modern plasma physics, has proposed that electrical currents flowing through the ionized plasma of space produce strong magnetic forces that greatly affect the development of galaxies.

The thin gas of space is ionized, producing what is called plasma. This means that electrons are separated from the nuclei of the atoms, and so can move freely through the plasma to form electric currents. Although the electric current flowing through a square meter of area is very small, the total current flowing through the huge area associated with a typical star can be very large. This area is several light years wide. The electric currents flowing through the ionized plasma of space can generate magnetic fields of very high energy, which interact with the magnetic field of the star and thereby alter the motion of the star. The electric and magnetic fields of the plasma currents tend to produce vortices that could cause the rotation and spiral shape of a galaxy. Over intergalactic distances, the electric and magnetic fields could create the spaghetti-like filaments into which galaxies are arranged.

2. The Big Bang Theory   27

Alfven has shown in laboratory experiments that instabilities cause plasma currents to form themselves into filaments, which are swirling electrical currents that twist relative to one another like the strands of a rope. Equivalent filaments are seen in the aurora, or "northern lights". They can also be seen in a gaseous nebula, which is a mass of heated plasma surrounding a group of stars. On an intergalactic scale, plasma electric currents could produce the filament arrangement of galaxies, and the larger ribbon-like structures.

Much more information is given by Lerner [4] concerning the tremendous cosmological implications of plasma physics, which has been pioneered by Nobel laureate Hannes Alfven.

The Big Bang cosmologists have ignored or dismissed plasma theory, and few have even bothered to read about it. The well-known cosmologist, P. James E. Peebles (whose *Scientific American* article was discussed earlier) stated that Alfven's ideas are *"just silly"*. His colleague at Princeton, Jeremiah Ostriker, commented, *"There is no observational evidence that I know of that indicates electric and magnetic forces are important on cosmological scales."*

Alfven, as well as lesser-known plasma physicists, have repeatedly had their papers rejected by astrophysical journals because they contradict Big Bang wisdom. Alfven commented, *"I think the Catholic Church was blamed too much for the case of Galileo — he was just a victim of peer review"*.

As evidence against the Big Bang has mounted, Big Bang cosmologists have shrugged it off, and have proceeded to devise more and more elaborate ad-hoc theories to bypass the evidence. Joseph Silk, who has written three books on the Big Bang, stated flatly

"It is impossible that the big bang is wrong. Perhaps we'll have to make it more complicated to cover the observations, but it is hard to think of what observations could refute the theory itself."

Lerner [4] (page 54) responded to this with the following, which must also have reflected the convictions of Nobel laureate Alfven, who helped Lerner greatly in his book:

"This attitude is not at all typical of the rest of science, or even of the rest of physics. In other branches of physics, the multiplication of unsupported entities to cover up a theory's failure would not be tolerated. The ability of a scientific theory to be refuted is the key criterion that distinguishes a science from metaphysics. If a theory

cannot be refuted, if there are no observations that could disprove it, then nothing can prove it — it cannot predict anything; it is a worthless myth."

Lerner (p. 56) quotes the following words by Alfven on the myth issue:

"The cosmology of today is based on the same mythological views as that of the medieval astronomers, not on the scientific traditions of Kepler and Galileo."

Big Bang theorists treat the Einstein theory as their ultimate truth (their infallible "Bible"). As reported by Lerner (p. 163), cosmology theorist George Field stated his Big Bang philosophy as follows:

"I believe the best method is to start with exact theories, like Einstein's, and derive results from them."

This philosophy of modern cosmology sharply contrasts with legitimate science, which demands consistency between theory and observation. It is the philosophy of mythology, not of science. Lerner (p. 162) states:

"Entire careers in cosmology have now been built on theories that have never been subject to observational tests, or have failed such tests and have been retained nonetheless."

As Lerner (p. 127) points out, the mythology of modern cosmology is based on the "myth of Einstein". He explains Alfven's concepts with:

"It is quite ironic that the triumph of science [from relativity theory] led to the resurgence of myth. The most unfortunate effect of the Einstein myth is the enshrinement of the belief, rejected for four hundred years, that science is incomprehensible, that only an initiated priesthood can fathom its mysteries."

Lerner quotes the following words of Alfven:

"The people were told that the true nature of the physical world could not be understood except by Einstein and a few other geniuses who were able to think in four dimensions. Science was something to believe in, not something that should be understood. Soon the best sellers among the popular science books became those that

2. The Big Bang Theory

presented scientific results as insults to common sense. One of the consequences was that the boundary between science and pseudoscience began to be erased. To most people, it was increasingly difficult to find any difference between science and science fiction."

Since the start of wide acceptance of the Big Bang theory about 1970, cosmology theorists have been struggling with its many serious conflicts with observational evidence. Lerner [4] (pps 150-163) gives an excellent discussion of the increasingly complicated, bizarre, and arbitrary hypotheses that have been developed since 1965 to accommodate the more and more evident conflicts of the Big Bang theory with observed data.

Quasar Studies by Astronomer Halton Arp

The extreme redshift of the quasar was discovered in 1963. A quasar is a star-like object, looking like a point of light to the telescope. It does not have an extended image like a galaxy. Unlike normal stars, the spectrum of a quasar has an extremely large redshift. If we assume that this redshift is caused by velocity, quasars must be receding from us at velocities approaching the speed of light, and so must be billions of light years away. If quasars are at such great distances, they must radiate enormous amounts of energy for us to see them.

A typical quasar appears to radiate more than ten times the total output from our Milky Way galaxy. However, the brightness of many quasars varies, over periods of months, weeks, days, and even hours. This variation indicates that a quasar has a size comparable to a star, not a galaxy. Some quasars, are no larger than our solar system. How can an object the size of a star radiate the energy of one trillion stars?

The quasar becomes a much more reasonable object if we assume that its redshift is caused primarily by gravity, rather than by velocity. In 1916, Einstein showed in his general theory of relativity that a gravitational field produces a spectral redshift. An important test used for verifying general relativity was to measure the redshift produced by the gravitational field of our sun.

Redshift is defined as the increase of wavelength divided by the normal wavelength. For example, if the measured wavelength is 1.5 times the normal wavelength, the increase of wavelength is 0.5 times the normal wavelength, and so the redshift is 0.5.

The maximum value of gravitational redshift that can be predicted by the Einstein theory is 2. Since quasars can have redshifts appreciably greater than 2, it is generally believed that the quasar redshift cannot be a gravitational effect. Besides, the Einstein theory has indicated that a star would exhibit strong oscillations unless its gravitational redshift is much less than 2. Such analyses have convinced most astronomers that the quasar redshift cannot be gravitational; it must be caused by velocity. With an extreme velocity, a quasar must be billions of light years away, and must radiate an unbelievable amount of power.

Big Bang cosmologists theorize that quasars close to us have burned themselves out long ago, because they consume energy at such a fantastic rate. We see the distant quasars only because we are looking at them as they were billions of years ago, due to the time it takes their light to reach us.

However, there is strong astronomical evidence that quasars are much closer than is commonly believed. Starting in the middle 1960's, the eminent astronomer, Dr. Halton Arp [5, 6], found strong observational evidence of quasars that are parts of nearby galaxies. He photographed three quasars that appear to be in the outer fringe of galaxy NGC 3842, yet have spectral redshifts far exceeding that of the galaxy. It is theoretically possible that this is a chance relationship, that the quasars actually lay far beyond the galaxy. However the probability that the images of three quasars would fall this close to an arbitrary direction in space is about one in a million.

Arp found several similar cases. *The possibility is essentially zero that all of these observations could be accidental, which show associations between quasars and nearby galaxies.* Besides this strong statistical evidence, Halton Arp also found several examples of filament structures that directly connect quasars to galaxies of much lower redshift.

The quasar observations by Arp were opposed by the general astronomical community, because they did not agree with the accepted dogma that quasars are billions of light years away. He found it very difficult to get his findings published. Some journals rejected his work, and papers were often held up for years by referees. Finally in 1984, the committee that controls observation time at the Palomar and Mount Wilson Observatories refused to allow Arp to use these facilities.

Dr. Halton Arp had performed distinguished research at Palomar and Mount Wilson Observatories since he received his PhD degree in 1953. He was president of the Astronomical Society of the Pacific from 1980 to 1983, and received awards from the American Astronomical Society, the American Association for the Advancement of Science, and the Alexander von Humbolt Senior Scientist Award. After being denied research facilities in California, he was forced to move to Germany to continue his career, where he joined the Max Planck Institute for Physics and Astrophysics in Munich.

The observations of quasars and the experiences by Arp are documented in the excellent books by Halton Arp [6. 5], *Quasars, Redshifts and Controversies*, and *Seeing Red: Redshifts, Cosmology, and Academic Science*. These books are essential reading by anyone who, like the writer, believes that objectivity in science is a crucial element of our society that must be preserved.

The traumatic experience of Halton Arp was summarized as follows by Fred Hoyle, Geoffrey Burbidge, and Jayant V. Narlikar [13] in their book published in the year 2000, *A Different Approach to Cosmology*:

> *"Arp's own colleagues at the Mount Wilson and Palomar Observatories . . . recommended to the directors of the two observatories that his observational program should be stopped, i.e., that he should not be given observing time on the [telescopes in these observatories] to carry on with this program. Despite his protests, the recommendation was implemented, and after his appeals to the trustees of the Carnegie Institution were turned down, he took early retirement and moved to Germany where he now resides, working at the Max-Planck-Institut fur Physik und Astrophysik in Munich. . . . Thus Arp was the subject of one of the most clear cut and successful attempts in modern times to block research which it was felt, correctly, would be revolutionary in its impact if it were to succeed."*

Halton Arp was not directly concerned with the Big Bang theory. However, if his observations were correct, the quasar redshift could not be explained by the Einstein theory, which Big Bang cosmologists use as their Bible. The general acceptance of the Arp discoveries would cast serious doubts on the validity of that Bible.

As we will see, the Yilmaz theory of gravitation has extended the Einstein theory and has corrected its weaknesses. With the Yilmaz theory, there is no fundamental limit to gravitational redshift, and so the Yilmaz theory can explain the redshifts of all quasars as gravitational effects. This result predicts that quasars are much closer than is commonly believed and are radiating much less energy. Therefore the Yilmaz theory agrees with the findings of Halton Arp. This issue is discussed further in Chapter 8.

Lack of Scientific Objectivity in Astronomy Today

As we marvel at the astounding advancements of scientific knowledge over the past 400 years, we should recognize that a fundamental requirement underlying the development of this knowledge is scientific objectivity. To achieve this objectivity, there must be open scientific debate. Without scientific debate, there can be no real science.

In Chapter 4 we will examine the evidence that has been amassed by paleontologists to develop our understanding of how life evolved on earth. This knowledge was obtained from a tremendous amount of painstaking work by paleontologists, carefully examining minute details in the fossils that have been found. To achieve reliable means of interpreting these findings has required a tremendous amount of scientific debate.

One example of this is the question of the ancestry of birds. Did birds evolve from dinosaurs, or from some other reptile? This relatively minor question has been heatedly debated for years. It is only with such debates that reliable conclusions can be realized.

Astronomy was originally a very objective science. However, since the rapid growth of the Big Bang theory in the 1960's, it has become highly dogmatic. Scientific facts are now forced to fit into a rigid mold. Astronomical evidence that does not agree with accepted dogma is ignored. We can be sure that many other scientists working in astronomy have learned by the example of Halton Arp, and are careful to find only those results that are consistent with the accepted concepts.

What is at issue is not a matter of whether the Big Bang theory is right or wrong. An honest assessment will clearly show that there are serious doubts concerning the theory. Yet the theory is treated as fact, not theory. Evidence that does not agree with the theory is suppressed. As a result, astronomy today is deteriorating into mythology. Conclusions are based on consensus rather than on evidence.

How does one find the truth? One must search for the facts, and let the facts speak for themselves. The reader should use his or her own intelligence to evaluate the evidence, and not be intimidated by the statements made by authorities. The statements of a scientific authority in this confused field of astronomy today may be influenced far more by economic and personal factors than by scientific evidence.

Chapter 3

Alternatives to Explain the Hubble Redshift

This chapter examines the major alternatives to the Big Bang theory that have been proposed to explain the Hubble redshift. As we will see, all of these cosmology theories have serious weaknesses.

This chapter also introduces a revolutionary new explanation of cosmology based on the Yilmaz theory of gravity. Is it correct? It certainly deserves serious consideration, if only because it adopts a radically new approach to the problem.

Well-Known Alternatives to the Big Bang

Steady-State Universe Theory

The *Steady-State Universe* theory was presented in 1948 by Fred Hoyle, Hermann Bondi, and Thomas Gold. It was based on the concept that the universe should look approximately the same to observers at any location, and at any point in time. To achieve this condition as the universe expands, matter is created to compensate for the Hubble expansion of the universe. They estimated that the required generation of matter would be equivalent to the creation of one hydrogen atom every year within a cubic volume that is 100 meters on a side. This would be far too small to be observed directly.

As explained by Lerner [4] (p. 145), the new matter that is created would appear in the form of hydrogen atoms. These gradually condense by their own gravity into huge clouds, then into galaxies, and finally into stars. The stars process the hydrogen atoms to form the heavier elements. As the fuel of a galaxy is exhausted over billions of years, the galaxy gradually dies, its stars becoming invisible dark embers. In the meantime, new galaxies come into being from the newly created matter.

A problem with the Steady State Universe theory is that it does not explain what is causing this spontaneous generation of matter. Is the matter created out of nothing? The Big Bang theory postulates an enormous creation of matter, which occurs all at once when the universe is born. To assume that matter is being created slowly out of nothing, as required in the Steady-State Universe theory, does not seem to be worse that assuming that it is created all at once out of nothing.

When the cosmic microwave background radiation was discovered by Penzias and Wilson in 1965, the Steady-State Universe theory did not have an immediate explanation for it, and so the theory was rapidly brushed aside by the momentum toward the Big Bang theory. This event is recounted in the recent book by Hoyle, Burbidge and Narlikar [13] (p. 84-85). The book concludes that the primary impetus for this change was the enormous expansion of funds for astronomy that resulted from the NASA program. The book reads (p. 84):

"[The 1965 discovery by Penzias and Wilson] coincided with the beginning of a great expansion of the number working in astronomy, an expansion begun by the intense interest at the time in space, which led to NASA being funded in billions of US dollars, rather than in the few tens of millions that had been previously available to astronomy, even in the most favorable circumstances. The number of attendees at the International Astronomical Union (IAU) jumped in only a few years from 300 or 400 to more than 2000, an immense expansion that made it possible to claim that cosmology as a branch of physical science only began in earnest in 1965."

Thus, Hoyle, et al, recognize that the stampede to the Big Bang, and the eclipse of the Steady-State Universe theory, was primarily the result of economic forces that happened to coincide with the discovery of the cosmic microwave background radiation by Penzias and Wilson. As this book has explained, another major factor was the wide availability of computers at that same time. Many theoreticians were induced to perform computer studies of the Einstein general theory of relativity, which led inescapably to the Big Bang theory.

When support for the Steady-State Universe theory waned, Bondi and Gold lost interest. With time, Hoyle gradually abandoned the theory and turned to another concept. Nevertheless, the theory was not disproved. Although it has fundamental problems, its weaknesses are certainly no more severe than those of the Big Bang theory.

Quasi Steady-State Cosmology Model

While not disavowing the Steady-State Universe theory, Fred Hoyle has turned his attention to a new concept, which he calls the Quasi-Steady-State Cosmology model, abbreviated QSSC. He is supported in this effort by Geoffrey Burbidge and Jayant V. Narlikar. The theory is described in their recent book, *A Different Approach to Cosmology* [13]. The book does an excellent job of documenting the many astronomical contradictions that have occurred in cosmological research.

The Quasi-Steady-State Cosmology (QSSC) model is actually more similar to the Big Bang theory than to the Steady-State Universe theory. It assumes that the size of the universe oscillates with a period of about 100 billion years. Rather than starting at an infinitesimal size, as postulated in the Big Bang theory, the universe is assumed to oscillate in size by a factor of 12. The present universe has about 6 times the minimum size, and half the maximum.

According to this theory, the diameter of the universe changes with time like the trajectory of a bouncing ball. The minimum diameter of the universe is not zero; it is 1/12 of the maximum. We are now on the rising part of the ball trajectory, about 10 billion years after the minimum point. In 40 billion years the ball will reach its maximum height and will begin to fall again, which means that the universe will stop expanding and start contracting. In 90 billion years, the ball will reach the minimum point, which means that the universe will reach its minimum size. Then, suddenly, the ball hits the floor and rebounds, which means that the universe suddenly changes from contraction to expansion.

The major question about the theory is, "What causes the universe to switch from contraction to expansion when the ball hits the floor?" The answer (p. 227) appears to be as follows:

"We have taken the commonsense alternative of supposing that energy appears in the universe in compensating positive and negative forms"... "Negative energy fields are inherently explosive. Concentrated locally they produce violent events, offering a possibility of explaining a wide range of observations --- radio galaxies, [quasars], and active galactic nuclei. When uniformly distributed, a negative energy field exerts a negative pressure, which shows itself in our view in the expansion of the universe. So far our discussion has been concerned with the last feature, namely the possibility of explaining the observed expansion of the universe

without recourse to a primordial explosion, the expansion of the universe being necessarily concomitant on the appearance of compensating positive and negative forms of energy as an ongoing phenomenon."

These authors are led to this oscillating universe theory in order to avoid the Big Bang singularity. When Big Bang theorists use the Einstein theory to extrapolate the universe expansion backward in time, they are led to the concept that the universe started as a singularity having practically zero size and infinite density of matter. The QSSC theory is being proposed to avoid this physically impossible condition. However, the concept of alternating positive and negative energy fields does not seem to be physically possible either.

An important aspect of this QSSC theory is that, like the Big Bang theory, it is also based on the Einstein general theory of relativity. In fact the original Steady-State Universe theory also applied the equations of the Einstein theory.

Variable Mass Theory

In 1964, Fred Hoyle and Jayant V. Narlikar proposed a gravitational theory (the Hoyle-Narlikar or HN theory) based on the interaction of particle masses, in which particle interactions propagate at the speed of light. The HN theory allows the masses of elementary particles (protons and electrons) to vary with time. If particle masses are constant, the HN theory reduces to the Einstein theory. When applied to cosmology, the Einstein theory naturally leads to a universe beginning as a Big Bang singularity having infinite density of matter. By allowing particle masses to vary, the HN theory can allow the universe to begin in a state of zero mass, and so can avoid the Big Bang singularity.

The matter in distant galaxies should be younger than matter on earth, because we are looking backward in time when we observe a distant galaxy. If the mass of an elementary particle increases with time, younger particles should generate radiation of longer wavelength than older particles. The reason is that the wavelength of a photon varies inversely with the mass of the electron that makes a transition. Hence the light from a distant galaxy should be of longer wavelength than on earth. This effect may explain the Hubble redshift.

In 1993, Narlikar and Arp [21] used the HN theory to explain the excess redshifts of quasars and galaxies that are unrelated to distance. They postulate that elementary particles of lower mass are continually being created, which form galaxies and quasars that emit longer wavelengths, and so have excess redshifts. However, if this postulate is true, we should expect to find protons and electrons on earth having different masses. There is no evidence of this.

Matter-Antimatter Theory

Lerner [4] (pp. 425-430) gives a thorough review of alternatives to the Big Bang theory in his Appendix A, "What Causes the Hubble Redshift". He starts with a critical evaluation of the Matter-Antimatter theory of Hannes Alfven, which is also discussed in his Chapter 6. This theory applies the principle of antimatter. Antimatter is similar to regular matter, except that it has the opposite charge. For example, the antimatter for an electron, called a positron, has the same mass as an electron, but has a positive charge. When elements of matter and antimatter collide, they annihilate one another and release pure energy. Antimatter has been created in the laboratory in very small quantities. When matter is created from energy in the laboratory, equal quantities of matter and antimatter are produced.

Alfven proposed that separate layers of matter and antimatter originally existed in the universe, while the universe was collapsing. At a crucial point, the separate layers met one another, causing a great explosion, which made the universe suddenly expand. The effect was similar to the bouncing ball effect proposed by the Quasi-Steady-State Cosmology theory.

Eric Lerner gives a detailed evaluation of the Alfven theory [4] (p. 425-428). His conclusion is, "The theory faces some definite problems, at least one of which appears, in my own view, to be fairly serious but not insurmountable." This is the assessment given by one who is personally biased toward the theory.

Photon Collisions with Hydrogen Molecules

This theory by Paul Marmet proposes that the Hubble redshift is caused by photon collisions with hydrogen molecules. This theory is also supported by Grote Reber, a pioneer in the development of radio astronomy. The theory is discussed by Lerner [4] (p. 428-429), and is analyzed in Appendix G of the website *Addendum* [2]. Paul Marmet [18] explains this theory in his own website.

The Marmet cosmology theory assumes that the Hubble redshift does not represent an expansion of the universe. It is a redshift effect caused when light passes through a very thin gas. Marmet's studies show that a photon gives up a tiny amount of energy when it collides with a hydrogen molecule, but its direction does not change. This loss of energy produces a redshift. Lerner notes that, "Marmet has calculated this [redshift] effect for our own sun, showing that it explains a long mysterious redshift between the limb and the center of the sun."

The Marmet redshift from one photon collision is equivalent to a Doppler velocity shift of 2 mm/sec. The number of collisions is proportional to the density of the hydrogen gas and the length of the path. For a density of one hydrogen molecule per cubic centimeter, there would be 300 collisions per light year of path length. Combining this with the previous number shows that the redshift over this one-light-year path length would be equivalent to a velocity of 600 mm/sec. Therefore the redshift over a path of one million light years would be equivalent to a velocity of 600 km/sec.

Thus, a gas density of one hydrogen atom per cubic centimeter throughout our universe would produce a Hubble constant of 600 km/sec per million light years. To achieve the Hubble constant that is commonly assumed today, 20 km/sec per million light years, the required density of gas would be 1/30 of a hydrogen atom per cubic centimeter. This is equivalent to 33,000 hydrogen atoms per cubic meter.

Lerner [4] (p. 428) reports that the density of luminous matter in the universe is about 0.1 hydrogen atom per cubic meter. Even with much more dark matter than has been found, the average density of the universe is not expected to exceed 10 hydrogen atoms per cubic meter. The density required by the Marmet theory is 3300 times greater than this.

Lerner concludes that the Marmet redshift effect is not sufficient to explain the Hubble redshift. However, Lerner states that, "Potentially this effect could explain the high redshifts of some quasars".

The writer agrees with Lerner. We will see in Chapter 8 that the Marmet redshift effect gives a promising explanation for the excessive redshifts of certain galaxies and may be an important factor in the redshifts of quasars.

Other Hypotheses for Explaining the Hubble Redshift

Lerner mentions the hypothesis by J. P. Vigier, which postulates a new term in the equations of quantum mechanics that causes the vacuum itself to absorb energy. Since this approach involves an unsubstantiated ad-hoc assumption, it is difficult to defend.

Lerner gives the following postulate made by quantum physics pioneer Paul Dirac in 1938: "Dirac proposed that instead of space between the galaxies expanding, . . . all space is expanding because the basic scale of all objects, from electrons to galaxy clusters, grows with time, due to an unknown physical law." Other versions of these concepts have been proposed.

Lerner concludes with the following, "Advocates of all of these hypotheses need to make far more precise predictions before they can be thoroughly tested and confirmed or refuted."

Summary of Well-Known Alternatives to the Big Bang

One of the reasons that the Big Bang theory has so much momentum is that its well-known alternatives all have serious weaknesses. The fallacy behind the Big Bang movement is not a matter of whether the theory is stronger or weaker than its competitors. The problem is that the Big Bang theory is being treated as fact, and research in astronomy is tightly constrained by that premise. Astronomical research is controlled so that only those studies that agree with the accepted dogma are publicized and funded.

This book presents a radically new cosmology theory for explaining the Hubble redshift, which has a much more solid scientific foundation than the Big Bang theory or any of its well-known alternatives. Let us examine this new approach to cosmology.

The Yilmaz Gravitational Theory

Development of the Yilmaz Theory of Gravity

Although the Big Bang theory, the Steady-State Universe theory, and the Quasi-Steady-State Cosmology theory are superficially different, they have one fundamental element in common. They are all based on the Einstein general theory of relativity. The Variable Mass theory is based on the Hoyle-Narlikar (HN) gravitational theory, which has limitations that are similar to the Einstein theory.

We will see in Chapter 10 that the Einstein theory cannot achieve a multi-body solution. Since the HN theory reduces to the Einstein theory when particle masses are constant, the HN theory should have this same limitation.

Computer studies have shown that the Einstein general relativity theory predicts a black hole singularity. Nevertheless, Einstein strongly opposed that concept because it drastically violates our laws of physics. This non-physical singularity suggests that there is something fundamentally wrong with the Einstein theory. The Hoyle-Narlikar theory also allows a physically impossible black hole, but the Yilmaz theory does not.

The Einstein general relativity theory was a monumental advance in our understanding of physics. This theory comprises a large body of sound scientific principles that Einstein developed in order to generalize his relativity concept. The theory is specified by its gravitational field equation, which is a complicated tensor formula that actually represents ten independent equations. This book will show that the Einstein gravitational field equation has a flaw. Many scientists have questioned the validity of this equation, because it was derived by Einstein in a somewhat intuitive manner.

Huseyin Yilmaz developed his theory of gravity while performing PhD research at the Massachusetts Institute of Technology in the early 1950's. He retraced the steps that Einstein took in developing general relativity, and discovered that Einstein had made an unnecessary approximation when calculating the wavelength change that is produced by gravity. Yilmaz applied an exact calculation to this step, and this resulted in the basic formula for his gravitational theory. Thus the Yilmaz theory is a direct extension of the Einstein theory.

Yilmaz mailed a copy of his findings to Einstein, but Einstein was too sick to read it. Albert Einstein died a short time later. Yilmaz published his basic theory in 1958 in the *Physical Review*. Since then Yilmaz has performed numerous studies to extend his theory, which have been published in prestigious scientific journals.

Yilmaz has proven that the Einstein gravitational field equation is not a rigorous solution to the principles of relativity. He has refined the Einstein theory to correct its weaknesses, and has thereby developed the rigorous solution to relativity that Einstein sought.

A primary fallacy underlying Big Bang research is that it has treated the Einstein gravitational field equation as absolute truth. This has resulted in numerous predictions that strongly conflict with scientific evidence.

Simplicity of the Yilmaz Theory

The Einstein theory is applied by solving its very complicated tensor gravitational field equation, which can result in millions of terms except for simple cases. When computers became generally available in the

1960's, these equations could be programmed into a computer and thereby solved for complex physical models.

In a computer program, the Einstein gravitational field equation is solved backward by applying an iterative computation procedure, which may involve many billions of computations. Sophisticated computer techniques have been developed to allow the iterative computer programs to converge to solutions. Since the mid 1960's, hundreds of scientists have developed elaborate mathematical procedures to implement computer solutions of the Einstein theory.

Like the Einstein theory, the Yilmaz theory also has a gravitational field equation. This equation has an additional term not included in the Einstein theory, and so is even more complicated. Nevertheless, the Yilmaz theory is very much easier to solve, because Yilmaz has derived a general solution to his gravitational field equation. With the Yilmaz theory, one does not solve the gravitational field equation when applying the theory. The fact that the Yilmaz theory has achieved this general solution proves that the theory has profound mathematical integrity.

With its great simplicity, one might think that the Yilmaz theory would be strongly applauded. This might be true if the Einstein theory were being applied to practical applications, but that is rarely the case. In the artificial Big Bang computer studies, designed to achieve academic prestige rather than practical results, the simplicity of the Yilmaz theory is a severe liability. With the Yilmaz theory, the sophisticated computer programs developed for the Einstein theory become obsolete. Since the 1960's, these programs have been developed with enormous difficulty by hundreds of scientists. Acceptance of the Yilmaz theory would destroy the gargantuan Big-Bang research effort, and would eliminate the need for its sophisticated computer programs.

Failure of the Einstein Theory in Cosmology

Yilmaz has demonstrated that the Einstein theory has serious theoretical weaknesses, which we will discuss later. Because of these weaknesses, the Einstein theory cannot be used to study cosmology. It can be applied reliably within our solar system, where gravitational fields are weak. However, where gravitational fields are strong, as they are in cosmology, the Einstein theory cannot yield meaningful predictions. The physically impossible black-hole and Big-Bang singularities are consequences of this failure of the Einstein's theory in intense gravitational fields.

Since 1929, countless scientists have struggled to develop an explanation of cosmology that is consistent with the Hubble redshift. In order to obtain a mathematical foundation for such a study, one needs a theory of gravity, and the Einstein general theory of relativity is usually employed for this purpose. However, the Einstein theory cannot yield reliable cosmological predictions.

An essential requirement for developing a meaningful approach to cosmology is a theory of gravity that works, and the Yilmaz theory provides it. This is the key for unlocking the mysteries of the universe. Let us see what the Yilmaz theory tells us.

The Yilmaz Cosmology Model

Basic Predictions of the Cosmology Model

When Yilmaz presented the first paper on his theory in 1958, he applied it to a simple cosmological concept, which we call the Yilmaz cosmology model. This model assumes a constant average density of matter, which extends to infinity and does not vary with time. To his surprise, Yilmaz found that his model predicts an expanding universe. Relativistic effects due to gravity cause the universe to expand, just as Hubble had observed.

How can gravity, which normally causes bodies to attract one another, force the universe to expand? At first sight, this may seem self-contradictory. However, we will see that it has a reasonable physical explanation.

Since the Yilmaz cosmology model assumes a constant average density of matter that does not vary with time, the expansion of the universe predicted by the model implies that matter must be created to compensate for the universe expansion, just as the Steady-State Universe theory postulates.

After writing his first paper, Prof. Yilmaz came to realize that cosmology can be very speculative, and so he has ignored cosmology in subsequent papers. He does not want the speculative aspects of cosmology to obscure the rigorous mathematical character of his relativistic theory. The writer is independently exploring the cosmology implications of the powerful Yilmaz theory of gravity.

When the writer extended the analysis of the Yilmaz cosmology model, he found the Hubble expansion to be a local effect, occurring within five billion light years. Over much larger distances the universe does not expand. This leads to the remarkable conclusion that the universe expands locally about every point of the universe, yet the overall size of the universe does not change. Although this concept may seem unbelievable, it is a direct consequence of the relativistic equations of the Yilmaz theory.

The Yilmaz cosmology model yields a radically new concept of cosmology. Is it correct? It is clear that none of the other explanations of cosmology provide acceptable explanations for the Hubble redshift. A new approach is obviously needed. The Yilmaz cosmology model satisfies the fundamental requirement that all other cosmology theories lack. It is based on a rigorous gravitational theory, and so its predictions should be taken seriously.

Cosmic Microwave Blackbody Radiation

As will be shown, the Yilmaz cosmology model predicts that Doppler-shifted light from distant galaxies should produce cosmic radiation at microwave frequencies. This predicted cosmic microwave radiation agrees closely in spectrum and intensity with the data measured by the COBE satellite. The radiation received from distant galaxies is strongly Doppler shifted, because the apparent velocity of a distant galaxy is very close to the apparent speed of light. Nevertheless the actual velocities of distant galaxies are small, because the apparent speed of light at that point is small. These relativistic effects are explained in Chapter 11.

The calculations show that the light from distant galaxies should produce cosmic microwave radiation that corresponds in intensity and spectrum to a blackbody at a temperature between 2.1 and 3.4 degrees Kelvin. This result agrees closely with the blackbody temperature of 2.73 degrees Kelvin that was measured for cosmic radiation by the COBE satellite. In agreement with the COBE data, the radiation predicted by the Yilmaz Cosmology Model should emanate uniformly from all directions, and its spectrum should closely follow that of an ideal blackbody.

Picture of the Universe Derived from Yilmaz Cosmology Model

Let us briefly examine the universe that is predicted by the Yilmaz Cosmology Model. The model predicts that relativistic effects due to gravity force the universe to expand locally. Since the model requires a constant average density of matter throughout the universe, matter must be created to compensate for the Hubble expansion. This creation of matter is consistent with our laws of physics, because the created matter

44 *How Was Our Universe Created?*

can be derived from energy that is radiated from stars in other parts of the universe. Einstein showed that matter and energy are equivalent.

The diffuse matter created in space congregates into gaseous clouds, which coalesce to form stars and galaxies. Nuclear fusion in the stars converts hydrogen into helium and then into carbon. This reduces the mass slightly, and the loss of mass is converted into a huge amount of energy. The released energy is radiated from the star as light and other electromagnetic waves, such as X-rays, radio waves, and infrared radiation. When the nuclear fuel is exhausted, most stars contract to become very dense stars called white dwarfs, which glow white hot from the gravitational energy that is released by the contraction. When a white dwarf star can contract no further, it cools to become a cold dark body, called a black dwarf.

We postulate that the energy from light and other electromagnetic waves that are radiated from a star is converted into diffuse matter in space. After the star stops radiating light and becomes a black dwarf, we postulate that the very dense star continues to radiate energy in the form of gravitational waves, until the mass of the black dwarf star is gradually dissipated. Gravitational waves are predicted by both the Einstein and Yilmaz theories. The energy of these gravitational waves is eventually converted into diffuse matter.

Our model predicts a universe that is continually changing, because of the conversion of matter into energy and energy into matter. Hence the universe appears to be eternally young although it is infinitely old.

This model of the universe is consistent with observational data. Although we do not know how the radiated electromagnetic and gravitational energy is converted into matter, this basic postulate of the model is consistent with physical laws. Unlike the Big Bang theory and its well-known alternatives, this cosmological theory does not postulate effects that violate our laws of physics and conflict with observational evidence.

This is a brief summary of our new picture of the universe. Does it represent reality? It certainly has a much firmer scientific foundation that the Big Bang theory.

Let us now explore some related scientific issues. We start by examining a question that has much firmer answers. How did life develop here on earth?

Chapter 4

How Was Our World Created?

Humans have continually asked the questions, "How were we created?" and, "How was our world created?" These questions are so important to us that the Bible starts with its story of Creation. Now we ask the scientists for their answers.

(The first section of this chapter, "The development of life on earth" is a repetition of the first section of Chapter 2 in *Universe* [1], but the bibliographic references have been omitted. The reader should consult *Universe* [1] for bibliographic references in this section.)

The Development of Life on Earth

The history of the earth and the evolution of its life have been deciphered from rocks and the fossils they contain. How old is the earth? The oldest earth rocks solidified 4.3 billion years ago. A meteorite fell in Arizona that is 4.5 billion years old, and so the earth and other planets must be at least that old. Based on this and studies of the moon, scientists have generally concluded that the earth was formed as a molten body about 4.6 billion years ago.

Early Life

Water gathered in the oceans, and it is there that life probably developed. The water came partly from condensed steam, but meteorites may have delivered most of it to the earth. Although the crust of the earth solidified about 4.3 billion years ago, studies of moon rocks and craters have shown that until 3.8 billion years ago the earth was bombarded so heavily by meteorites it was a very hostile environment.

Biologists now classify all life forms into three categories: *bacteria* and *archea*, which are simple cells without nuclei, and *eukaryotes*, which are much larger and more complex cells that contain nuclei. All multi-celled organisms consist of eukaryote cells.

Archea were recently discovered in volcanic thermal vents on the ocean floor, and typically live at temperatures close to the boiling point. They were first thought to be bacteria, but DNA studies show they are radically different. Archea can derive nourishment by combining volcanic hydrogen and carbon dioxide to form methane and water.

The primary chemical building blocks of life are amino acids. The ocean-floor volcanic vents have the chemicals and energy that could allow the basic amino acids to be formed by inorganic means. Therefore many biologists now believe that archea were the first organisms on the earth. Archea could have flourished in the hostile environment 4.3 to 3.8 billion years ago, when the earth was heavily bombarded by meteorites.

The first clear evidence of life occurred 3.6 billion years ago in the form of structures called stromotolites produced by colonies of *cyanobacteria*. The early organisms were bacteria and archea, which lack nuclei. It took nearly one billion years before complex eukaryote cells containing nuclei appeared in the fossil record, 2.7 billion years ago. Fossils of eukaryote cells can be identified because these cells are much larger than bacteria and archea, and they contain different biochemicals.

Photosynthesis is the process operating in cyanobacteria, algae, and terrestrial plants that uses energy from sunlight to produce food. Carbon dioxide from the air and hydrogen from the water are combined to synthesize carbohydrates. Oxygen is released into the air as a byproduct. Photosynthesis is implemented by chlorophyll, which looks green because it absorbs red and blue-violet wavelengths. Terrestrial plants are usually green, but seaweed (also called algae) can have different colors because other pigments mask the green chlorophyll color.

Photosynthesis was first performed at least 3.6 billion years ago by cyanobacteria, which are also called blue-green algae. These are simple bacteria cells without nuclei. Single-cell eukaryote *algae* (with nuclei) probably appeared 2.7 billion years ago. Multi-celled algae, or seaweed, appeared about 1.8 billion years ago. Oxygen was produced by the photosynthesis performed by cyanobacteria and algae. The oxygen gathered in the atmosphere and finally reached sufficient concentration to support complex animal life.

4. How Was Our World Created? 47

The First Animals

Abundant fossils of multi-celled animals suddenly appeared in our oceans at the start of the *Cambrian* period, 543 million years ago. Within 5 to 10 million years an evolutionary explosion occurred, in which animals from essentially all of the animal types (or phyla) have been identified. This includes a specimen from our own phylum, the *chordates*, which consist primarily of *vertebrates* (animals with backbones). Early chordates were nearly brainless filter feeders, but had a notochord nerve trunk along the body that was the predecessor of the spinal cord.

Scientists have long been amazed by the extremely fast evolutionary pace at the start of the Cambrian, and are attempting to explain it. Part of the answer lies in the *Ediacara* (or *Vendian*) period, which preceded the Cambrian. Fossils are scarce in the Ediacara, but this period (about 60 million years long) had many soft-bodied organisms, which left few fossils. Some of these seem to be predecessors of mollusks, crustaceans, and earthworms. However many were strange organisms, like small fluid-filled air mattresses, unlike any plant or animal today.

New insight into the ultra-fast Cambrian evolution has come from very recent studies showing that the earth was hit by as many as four extreme ice ages in the period between 750 and 580 million years ago. The following theory is gaining acceptance. Since snow is white, it reflects most of the sunlight back into space, and tends to make the earth colder. If glaciers get large enough, the cooling process can feed on itself, and the whole earth can freeze. Geological evidence appears to show that extreme ice ages occurred in which the average earth temperature dropped to -50 °C (-60 °F). The oceans froze to a depth of half a mile and stayed in this condition for 10 million years.

Volcanoes are a rich source of carbon dioxide. During rain, the carbon dioxide in the air reacts with silicates and carbonates in rock to form soluble bicarbonate compounds and silicon dioxide. This process removes much of the carbon dioxide from the atmosphere. However there was no rain during these extreme ice ages, and so a large concentration of carbon dioxide from volcanoes could accumulate in the atmosphere. This produced a greenhouse effect that finally melted the frozen earth, and soon created a very hot earth. Temperatures rose until the average earth temperature reached 50 °C (120 °F). After that, the temperature gradually declined and a normal climate returned.

The severe ice ages between 750 and 580 million years ago probably killed most of the life on earth. However marine life may have survived in regions around undersea volcanic vents. These vents could have melted chimneys through the ice sheets, leaving isolated ponds of clear ocean water. In these isolated pockets of life, different forms of animal life could have developed, which may have created many phyla of animal life that became evident when the Cambrian began.

This theory of the Precambrian ice ages is only a few years old. It will be studied and debated for many years before it can gain general acceptance. Nevertheless, it is clear that extreme climates occurred between 750 and 580 million years ago. These extreme climactic conditions probably had a strong influence on the evolutionary explosion of animal life that occurred at the start of the Cambrian period, 543 million years ago.

Development of Fishes

Marine animals evolved rapidly after the Cambrian explosion. The first vertebrate appeared 530 million years ago, and was a jawless fish, like a modern-day lamprey. Fishes with jaws appeared 30 million years later. Most of the early fishes were protected by armor plates, but armor gradually disappeared. The ancestors of sharks and rays (which have skeletons of cartilage) appeared about 430 million years ago, and modern bony fishes appeared about 390 million years ago.

Cartilage is lighter than bone, and so the cartilage skeleton allows a shark or ray to be only slightly heavier than the water that it displaces. Consequently these fishes can stay at a constant depth with only a slight upward force, which is achieved by a slow forward motion. Modern bony fishes are basically much heavier than the water that they displace, but they compensate for this with an air bladder. The fish extracts air from the water, and keeps its air bladder properly inflated to achieve neutral buoyancy. The air bladder allows a modern bony fish to stay motionless at a fixed depth.

The development of the air bladder in modern bony fishes about 390 million years ago was a key feature in fish evolution. Primitive fishes were probably much heavier than water, and lived near the bottom.

Amphibians Invade the Land

In the modern lungfish, the air bladder also acts as a lung. The lungfish can obtain oxygen either from the air (with its air bladder) or from the water (with its gills). An extinct relative of the lungfish evolved into an *amphibian*. Modern amphibians consist of frogs, toads, and

4. How Was Our World Created?   49

salamanders. They begin life in the water like fish, and develop lungs and feet as they mature.

Amphibians apparently evolved from fish that inhabited swamps. These fish had lungs so they could breath in oxygen-poor water, and had four feet-like fins to push them through shallow swamps. Some of these amphibian-like fishes actually had feet, and so were once thought to be true amphibians. However the feet were too weak to support the body on land, and could only have supported the body in water.

Amphibians moved onto the land about 350 million years ago. Initially they ate millipedes and similar bugs that fed on plants. Plants had spread over the land 80 million years earlier.

Spread of Plants over the Land

We have seen that multi-celled marine plants, which are called algae or seaweed, appeared about 1.8 billion years ago. Terrestrial plants evolved much later, because their physiology is more complicated. They require specialized structures to obtain and hold moisture and nutrients.

Primitive terrestrial plants, now extinct, spread onto the land 430 million years ago. These were followed by *mosses* and *liverworts*. Liverworts, which are more primitive than mosses, form small mats that sometimes resemble livers. Mosses and liverworts are low plants that require abundant moisture because they lack a vascular system. Next came the *ferns*, *club-mosses*, and *horsetails*, which have vascular systems for conducting fluids. These plants all reproduce by spores. Unlike a seed, which has many cells, a spore is a single cell.

Modern club-mosses are also called ground pines, because they often look like miniature pine trees. However many ancient club-mosses resembled palm trees, having grass-like leaves on top of a long trunk. Horsetails are hollow reeds, like bamboo stalks, that have leaves like pine needles growing at the joints between reed segments. Modern horsetails are typically 2 to 6 feet tall, but some ancient species were 60 feet tall and one foot in diameter.

In the *Carboniferous* period (360 to 286 million years ago), giant versions of ferns, club-mosses, and horsetails, 50 to 130 feet tall, formed great tropical jungles. These produced the massive coal deposits that we use today for fuel.

Cycads and *conifers*, which have seeds, appeared about 300 million years ago. Conifers bear cones, and consist of pine, spruce, cedar, fir, etc. Cycads resemble palm trees, but are biologically very different. Most cycads have become extinct, and only a few species survive today.

Flowering plants evolved much later, in the middle of the dinosaur period. The oldest definite fossils of flowering plants are 120 million years old, but flowering plants probably occurred early in the Cretaceous period, which started 146 million years ago. Birds evolved about the same time as flowers. What a coincidence that *flowers* and *birds*, which so enrich our lives with colorful beauty, were created about the same time!

Fungi, which consist of mushrooms, molds, yeast, etc., were originally classified as plants. However fungi are radically different from plants, which contain chlorophyll, and so are now placed in a separate kingdom. Like animals, fungi achieve nutrition by digesting carbohydrates, but fungi perform this digestion externally by excreting enzymes. Fungi reproduce by spores.

Lichens played a key role in the spread of terrestrial plants over the land. Lichen is a symbiotic combination of fungus and algae (or cyanobacteria). The fungus forms a body that gathers and holds water and nutrients, and the algae perform photosynthesis to produce carbohydrate food. Lichens can grow on bare rock. They gradually dissolve the rock, converting it into soil on which terrestrial plants can grow. The earliest definite fossils of lichens are 400 million years old. Lichens probably existed before that, but they live in conditions where fossilization is rare, and so the lichen fossil record is meager.

As explained earlier, carbon dioxide in rain converts the carbonates in rock into soluble compounds. This process breaks down rocks and thereby helps to form soil in which plants can grow.

Single-celled eukaryotes (cells with nuclei) are called *protists*, and first appeared 2.7 billion years ago. Protists consist of plant-like cells (such as diatoms), animal-like cells (such as protozoa and amoebae), and fungus-like cells. Plant-like protists are also called algae.

The Reign of the Reptiles

As stated earlier, the first amphibians moved onto the land 350 million years ago. Fifteen million years later the first *reptile* appeared. Reptiles have amniotic eggs similar to bird eggs, which allow them to reproduce on land. Amphibians, such as frogs, have small fragile eggs, like fish eggs, which must hatch in water. Amphibians dominated the land for 35 million years until they were displaced by reptiles.

Within 10 million years after the first reptile appeared, the reptiles separated into two groups: (1) the *Synapsid reptiles*, which evolved into mammals, and (2) the *Diapsid reptiles*, which evolved into lizards, crocodiles, dinosaurs, and birds. It is often thought that a third group, the *Anapsids*, evolved into turtles, but turtles probably descended from Diapsids.

The Synapsid reptiles, which are the ancestors of mammals, quickly became the dominant reptiles. They controlled the land for 70 million years, from 315 million years ago until the end of the *Permian* period, 245 million years age. At that time a mass extinction occurred. This was by far the most catastrophic extinction since the Cambrian began, and it eliminated 95 percent of all animal species.

Some of the synapsid reptiles survived the Permian extinction, but in the following *Triassic* period the synapsids were gradually eclipsed by a group of Diapsid reptiles, called Archosaurs, which includes crocodiles and *dinosaurs*.

A small mass extinction occurred 225 million years ago, and after that the first dinosaur appeared. A much more severe mass extinction occurred 208 million years ago, at the end of the *Triassic*.

The causes of the mass extinctions ending the Permian and Triassic periods (245 and 208 million years ago) are being studied extensively. Manicouagan Lake in Quebec, Canada lies in a 100-km wide meteorite crater, formed between 206 and 214 million years ago. This impact from an asteroid or comet may have caused the Triassic extinction 208 million years ago. However, new studies relate these extinctions at the ends of the Permian and Triassic periods primarily to enormous volcanic activity. Volcanoes released high levels of carbon dioxide to produce a greenhouse effect that caused severe global warming. The high temperatures could have devastated plant and animal life. The high carbon dioxide level in the ocean at the end of the Permian was augmented by radical changes in ocean currents caused by continental drift.

Although the causes of the mass extinctions at the ends of the Permian and Triassic periods will be strongly debated for some time, the consequences of these extinction events are clear. They radically influenced the evolutionary development of dinosaurs and mammals.

The dinosaur movie *Jurassic Park* has popularized the name of the *Jurassic* period, which followed the Triassic and began 208 million years ago. For more than 140 million years, in the Jurassic and the subsequent *Cretaceous* period, gigantic dinosaurs controlled the land. Our children know many of the dinosaur names. The largest were the house-size Brachiosaurus and Brontosaurus, also called Apatosaurus. The weirdest looking were Stegosaurus, with triangular bony plates along its back, and the tank-like Triceratops with three horns on its head.

By far the most ferocious predator was Tyrannosaurus Rex, which appeared in the Cretaceous period.

During the dinosaur era, the air was dominated by strange flying reptiles called Pterosaurs or Pterodactyls, which were closely related to dinosaurs. Some were small, but one Pterodactyl had a wingspan of 40 to 50 feet.

The seas contained many kinds of Diapsid reptiles. This included the Plesiosaur (7 to 47 feet long), which had an oval body with four paddle-like flippers, and a neck so long it looked like a sea serpent. The Ichthyosaurs looked like giant fish, some being 50 feet long.

The first *bird* (Archaeopteryx) appeared in the middle of the dinosaur era, and apparently evolved from a dinosaur. The original function of feathers was probably for insulation. A recent fossil appears to show a dinosaur with feathers that could not fly.

The Slow Rise of the Mammals

Many of the Synapsid reptiles that survived the mid-Triassic extinction 225 million years ago were quite similar to mammals, and from these came the first true *mammal*, which appeared 215 million years ago. The mammals were small during the dinosaur era. Most were the size of a mouse and none was larger than a domestic cat. Mammals remained tiny for 150 million years until the dinosaurs were eliminated. Two key features of a mammal are accurate tooth occlusion (so that it can chew effectively) and the ability to chew and breathe at the same time.

Although the mammals evolved from Synapsid reptiles, the early Synapsid reptiles, and many of the Synapsids that became extinct in the Permian and early Triassic periods, did not look at all like mammals. An early Synapsid reptile called Dimetrodon was 10 feet long and looked like a lizard with a giant sail-like fin on top of its body. The Dimetrodon is often included in the set of "dinosaur" models made for children's toys. We know that the Synapsid reptile was related to the mammal because, like the mammal, each side of the skull has a single hole behind the eye socket. In contrast, a Diapsid reptile skull normally has 2 holes behind the eye socket.

Sixty-five million years ago, at the end of the Cretaceous period, the dinosaurs were suddenly driven into extinction. It is commonly believed that this was caused by an asteroid or comet, about 10 kilometers in diameter that hit the earth near the Yucatan peninsula in Mexico. This caused a conflagration that killed nearly all plant and animal life on land

and in the seas. The dinosaurs, pterosaurs, and marine reptiles became extinct. However enough of the small mammals and birds survived to continue their species, along with a few cold-blooded representatives of the once-dominant reptiles.

On the other hand, since 1978 Professor Dewey McLean of Virginia Polytechnic Institute has written extensively showing that enormous volcanic activity occurred at the end of the Cretaceous period. Volcanoes flooded a million square miles in India, and a pile of volcanic lava near Bombay is 1.5 miles thick. If volcanoes could have caused the massive extinction at the end of the Permian period, it seems clear that this volcanism should have strongly affected the dinosaur extinction. This suggests that two separate disasters may have contributed to the destruction of dinosaurs at the end of the Cretaceous period.

With the competition from the dinosaurs eliminated 65 million years ago, the mammals took over the land and evolved rapidly. The birds flourished also, filling ecological niches in the air previously dominated by pterosaurs. Mammals and birds moved to the seas, as whales, porpoises, seals, and penguins, taking the place of extinct marine reptiles. The mass extinction 65 million years ago ended the age of reptiles and ushered in our modern age of mammals.

Many dinosaurs at the end of the Cretaceous period were probably more intelligent then the modern reptiles (turtles, lizards, snakes, and crocodiles} that survived the mass extinction. Nevertheless the mammals were apparently much more intelligent than the dinosaurs that they replaced. It seems likely that the 150 million years of dominance of dinosaurs over mammals resulted in a large increase in mammalian intelligence. Brainpower was probably a great asset in helping a tiny mammal from being eaten by a terrifying carnivorous dinosaur.

The *primate* order consists of humans, apes, monkeys, and a primitive group that includes lemurs and tarsiers. The first primates appeared at the end of the Cretaceous period just before the dinosaurs were eliminated, 65 million years ago. The first ape appeared 25 million years ago. Apes differ from monkeys in that they have no tail, they have a much larger brain, and they move in an upright manner when in the trees. Monkeys scamper on four legs along tree branches. Modern apes consist of the gibbon and orangutan, which live in Asia, and the chimpanzee and gorilla, which live in Africa.

The Ascent of Humans

About 6 million years ago, an African ape, called *Australopithecus*, developed the ability to walk upright. Climate changes had thinned out the jungle trees, and this ape needed to move over appreciable distances between trees. Walking upright was a natural evolution for an ape, and it had the great advantage that the ape could carry clubs and rocks in its strong arms to defend itself against predators, as chimpanzees occasionally do today.

About 2.5 million years ago, the first member of our own genus, *Homo Habilis*, evolved from this upright ape. He had the intelligence to make crude stone tools. He had learned to hunt animals to enrich his diet, and the stone tools were used in butchering the animals. *Homo Erectus*, having a larger brain, evolved about 1.8 million years ago. He was tall (about 6 ft) and a good runner. Homo Erectus had the adaptability that allowed him to leave Africa and populate Europe and Asia. He eventually learned to use fire.

Homo Habilis made crude pebble tools (called *Oldowan*) by striking one lava pebble against another. More complicated stone tools (called *Acheulian*) were made by Homo Erectus..

Starting about 500 thousand years ago, more intelligent beings evolved from Homo Erectus, which are known as *Archaic Homo Sapiens*. Their brains were larger, and the types of stone tools increased. A more advanced stone tool technology (called *Mousterian*) appeared about 200 thousand years ago. *Neanderthal man*, who had a heavy frame and was physically much stronger than modern man, appeared at least 200 thousand years ago. He had a brain as large as modern man. He had a sloping forehead, a heavy brow ridge, a very large nose, a protruding jaw, and a receding chin. He was cold adapted and was able to survive in the cold regions of Europe until about 30 thousand years ago.

The first anatomically modern human (*Homo Sapiens Sapiens*) appeared about 100 thousand years ago. However it was not until 40 thousand years ago that sophisticated implements of stone and bone appeared, which demonstrated the high level of intelligence of modern humans.

Agriculture was first developed about 12 thousand years ago. It started with the planting of grains and the domestication of sheep, goats and pigs. With agriculture, the population density could increase greatly. About 10 thousand years ago, the first towns of appreciable size appeared. One of these was at the site of the Biblical town of Jericho.

There is healthy debate over the details of the evolution of life on earth, and of how humanity has developed. Nevertheless the general outline of the process seems to be reasonably well understood and widely accepted. Now let us turn to the broader question. How was our sun created, along with its solar system that includes our earth, and how was the whole universe created?

The Creation of Our Solar System

Along with other planets, the earth revolves in our solar system around the sun. Our sun is an average-size star within an aggregation of more than one hundred billion stars that form our *Milky Way* galaxy. Our galaxy is similar to the M51 spiral galaxy, called the *Whirlpool galaxy*, which was shown in Fig. 1-1 of Chapter 1. The Milky Way galaxy is 100 thousand light-years in diameter, and 3500 light-years thick. It takes 100 thousand years for light to travel across our galaxy. Our sun is located 2/3 of the distance from the center of the galaxy to the circumference, and lies within a spiral arm. The sun is revolving with the galaxy, and completes one revolution every 220 million years. Because of this galaxy rotation, the sun is moving at a velocity of 250 kilometers per second.

The name M51 for the galaxy in Fig. 1-1 is the designation given in Messier's *Catalogue of Nebulous Objects*, which Charles Messier (1730-1817) published in 1784. Stars look like points of light to a telescope, but "nebulae" have extended images. We now know that nebulae can be radically different from one another, and consist of galaxies, stellar clusters, and gaseous clouds. Messier developed his catalogue because he was looking for comets, and wanted to distinguish comets from fixed bodies. He discovered at least 15 comets, but this catalogue is what made the Messier name famous, and it serves as the primary reference for locating extended objects in the sky. Messier lost his astronomer job when the French Revolution erupted in 1789.

The M51 Whirlpool galaxy is 35 million light years away. Its spiral shape was first observed in 1845, and it was the first spiral "nebula" to be discovered. It was originally thought to be a solar system undergoing formation. It was not until the early 1900's that astronomers proved that M51 and other similar nebulae are galaxies containing billions of stars, like our own Milky Way galaxy. The M51 Whirlpool galaxy is located in the sky about 2 degrees southwest from the end of the Big Dipper handle.

By studying stars in various states of development, astronomers have been able to piece together the stages in which a star evolves. A

56 How Was Our Universe Created?

cloud of gas and dust is drawn together by the force of gravity. The energy released by gravity heats the gas until nuclear fusion occurs, and a star is formed. The energy from the fusion keeps the star from collapsing further. Stars of the general size of our sun reach a stable state, and radiate energy like our sun for many billions of years.

How was our solar system created? Angular momentum was an important factor in its formation. Although the sun has nearly 99.9 percent of the mass of the solar system, it has only 2 percent of the angular momentum. Jupiter has 70 percent of the angular momentum, and Saturn has nearly all of the rest. By considering the rotation of our galaxy, one can estimate the angular momentum of the original cloud of gas and dust that condensed to form our sun. That cloud should have had hundreds of times the angular momentum that our whole solar system has today. Therefore we ask, "How did our sun lose its angular momentum as the gas and dust cloud collapsed to form a star?" The answer to this question is a key aspect of the formation of our solar system.

As explained by Lerner [4] (pp. 188-189, 209), Hannes Alfven has proposed the following theory of the development of our solar system, which is based on plasma physics. Electric currents flow through the ionized gas in space. Over the very large regions of space, these currents are huge and generate enormous magnetic forces.

Since the whole galaxy is in rotation, a gas cloud is initially rotating. The angular rate of rotation increases as the gas cloud condenses to create a star, like a figure skater who spins faster by drawing the arms close to the body. The magnetic fields in the rapidly rotating star generate electric currents in the surrounding ionized gas. The magnetic forces generated by these currents transfer angular momentum from the star to the gas. By this means the star gradually loses angular momentum.

This process results in a rotating disk of gas and dust surrounding the star. The material in the disk gradually condenses into bodies to form a system of planets rotating around the star. Gravitational attraction causes dust to accumulate into grains, which grow into balls, etc. The largest body eventually sweeps up the rest within a range of distance from the star, until there are a limited number of planets revolving around the star. Plasma electric currents accelerate the condensation of the gas and dust into planets.

Regardless of the details of the process that created our solar system, angular momentum considerations indicate that the generation of a solar system around a star is probably a normal development. This conclusion is supported by recent astronomical observations of infrared stellar radiation, which show that stars less than 400 million years old are usually surrounded by dust clouds, but older stars are generally not. This suggests that planets formed around the older stars have swept up the dust clouds.

Since a solar system around a star is probably common, we can conservatively expect that at least a billion of the more than one hundred billion stars of our Milky Way galaxy should have planets like our earth that can support life. However such a conclusion has little practical meaning. It does not seem possible that we will ever be able to communicate with, much less visit, any planet of our galaxy that is not located close to us. The enormous distances across our galaxy should place all but the nearest stars beyond man's detailed knowledge.

Our sun is located about 30 thousand light years from the center of our galaxy. If we had a space ship that moved at the speed of light, which is more than 10 thousand times faster than any space vehicle built to date, we could have traveled only 40 percent of the distance to the galaxy center, if we had started when humans first began to use agriculture, 12 thousand years ago.

The nearest star, Proxima Centauri, is 4.3 light years away. It is 9000 time further than the planet Neptune, and is 100 million times further than the moon, which is the greatest distance yet traveled by humans. There are 62 stars within 17 light years.

The distances within our galaxy are vast, but they are infinitesimal in comparison to the dimensions of our observable universe.

The Hubble Expansion of Our Universe

Our Milky Way galaxy is bound together gravitationally with a group of galaxies and stellar clusters called the *Local Group*. The largest members besides our Milky Way galaxy are: M31 in the constellation Andromeda (2.2 MLy), M33 in the constellation Triangulum, which is next to Andromeda, (2.5 MLyr), and the Large and Small Magellanic Clouds (0.17 and 0.3 MLyr), which are visible only in the southern hemisphere. The values in parentheses give distances from earth in millions of light years (MLyr). The most distant member of this group is 5 million light years away. The Local Group does not expand by a measurable amount, but beyond this region all of the galaxies are moving away from us.

Edwin Hubble discovered that all galaxies beyond our Local Group are moving away from us at velocities approximately proportional to distance. This indicates that our universe is expanding. Based on recent data, the Hubble expansion rate is about 20 km/sec per million light years. If galaxy motions are extrapolated backward in time, using this Hubble expansion rate, the whole universe seems to have emerged from a single point 15 billion years ago.

A period of 15 billion years may seem infinitely long in comparison to recorded history. However, we have seen that the earth was formed 4.6 billion years ago, and the sun was created about 5 billion years ago. Do we really believe that our whole universe, containing many hundreds of billions of billions of stars, can be only 3 times as old as our sun?

This book gives a radically different interpretation of the Hubble expansion. The Yilmaz cosmology model predicts that the Hubble expansion is a relativistic effect that is caused by gravity, and that the age of the universe is infinite. The Hubble expansion is an essential characteristic of a universe that is infinitely old, because it produces the continual change that keeps the universe from dying

In order for the average density of matter to remain constant as the universe expands, matter must be created to compensate for the Hubble expansion. It is reasonable to assume that the new matter is derived from energy radiated from stars, and is created throughout space in the form of diffuse matter. This diffuse matter would eventually congregate to form new stars and galaxies, and the process would continue indefinitely.

Chapter 5

Discovery of the Nature of Light

The theory of relativity evolved from a study of the speed of light. This chapter reviews the history of our knowledge of light in order to introduce relativity theory and other principles that are needed in our investigation.

(This chapter is a repetition of Chapter 3 in *Universe* [1], but the bibliographic references have been omitted. The reader should consult *Universe* [1] for bibliographic references.)

Early Concepts of Light

Since the days of the Greeks we have known that sound is a vibration, or oscillation. One can feel the vibration in a musical instrument. The sound vibration travels as a wave through the device that generates the sound, and then through the air to the ear that receives it. Sound moves through the air as a compression wave. The air particles vibrate back and forth in the direction of propagation of the sound wave.

A sound wave can be visualized by dropping a pebble in the center of a smooth pond and watching the waves propagate outward over the surface of the water. The particles of water move very little; it is the energy of the wave that travels across the pond. Each particle of water vibrates up and down, and back and forth, following a small eliptical path.

But what is light? Is it a wave like sound? Or is it a stream of infinitesimal particles shot from the light source like stones from a slingshot?

Galileo, Copernicus, and Kepler

The science of optics started with the making of spectacles to correct visual defects. The first documented use of eyeglasses was a comment by Roger Bacon in 1268, but the Chinese may have used eyeglasses hundreds of years earlier. By 1629 eyeglass making was so advanced that a guild of spectacle makers was formed in England.

About 1608 a spectacle maker in Holland, probably Hans Lippershay, built the first telescope, which consisted of a convex objective lens and a concave eyepiece. (A convex lens is curved outward, whereas a concave lens is curved inward, like a *cave*.) Galileo, a professor in Italy whose full name was Galileo Galilei (1564-1642), heard of this and began making his own telescopes. In 1609 he supplied the governor *(doge)* of Venice with a telescope. This was so valuable for naval use that Galileo's salary was doubled, and he received lifelong tenure as a professor. Later in that year, Galileo built a 20-power telescope, which he directed at the heavens.

In 1610 Galileo revolutionized astronomy with his telescope. He studied our moon and found that it has mountains and valleys like the earth. He looked at Jupiter and made the fantastic discovery that Jupiter has moons of its own. He saw that Venus has phases like our moon, which showed that Venus must rotate around the sun. Galileo published his findings, and claimed that his astronomical observations proved that the *heliocentric* (sun-centered) theory of Copernicus must be correct. This theory conflicted with the strong belief in the *geocentric* (earth centered) concept of Aristotle and Ptolemy (c. 100-170 AD) that the sun and planets rotate around the earth.

In 1543 Nicholaus Copernicus (1473-1543), a Polish astronomer whose Polish name was Mikolai Kopernik, had advanced the revolutionary concept that the earth and planets rotate around the sun. He worked as clergyman (possibly a priest), secretary, and physician at the Frauenburg cathedral in Polish East Prussia. The manuscript describing his theory was completed by 1530 but was not published until 1543, just before he died. A preliminary version of the new theory had received harsh criticism from many sources, including Protestant leaders Luther and Calvin, but Pope Leo X had expressed open-minded interest. Copernicus withheld publication of his full document because he was afraid of the severe criticism it would receive. He agreed to let friends publish it when he knew he was close to death.

Another strong supporter of the Copernicus concept at the time of Galileo was Johannes Kepler (1571-1630), a German astronomer. In

5. Discovery of the Nature of Light   61

1601 Kepler became director of Tycho Brahe's observatory near Prague, after Brahe died. For many years, Tycho Brahe (1546-1601), a Danish astronomer, had made accurate measurements of planet and star locations using very large instruments having no optical magnification. Kepler applied Brahe's data to the Copernicus model, and derived three accurate relations for the orbits of the planets that are known as Kepler's Laws. These are:

(1) A planet orbits the sun in an elliptical orbit, with the sun at one focus of the ellipse.

(2) A planet moves more rapidly when nearer the sun than when further away, such that a radius drawn from the planet to the sun sweeps over an equal area for an equal time interval.

(3) The expression r^3/T^2 is the same for all planetary orbits, where r is the mean distance from the orbit to the sun, and T is the period of the orbit.

The major publication by Kepler was his *Epitome of Copernican Astronomy*, published in 1621. This was the first astronomical textbook based on the Copernican system, and was the primary source of information on the subject for 30 years. Kepler died in poverty in 1630. Because of war, Kepler was unable to collect the arrears of his Imperial salary. Another personal blow to Kepler was that his mother was imprisoned 13 months for witchcraft, and died in 1622 soon after her release.

Kepler invented the astronomical telescope, which uses a convex rather than a concave lens for the eyepiece. This telescope has a wider field of view, but has an inverted image and so is not good for terrestrial use. The inverted image is not a problem in astronomy. Modern high-quality terrestrial telescopes and binoculars use the Kepler telescope concept, along with an additional lens or prism to correct for the inverted image. Inexpensive *field-glass* binoculars use the simple Galileo telescope concept with a concave eyepiece.

In Italy Galileo received very strong criticism for his support of the Copernicus concept from a number of influential university professors and clergymen. Galileo contributed to this conflict by writing material that was taken as a personal affront by politically powerful intellectuals. They finally convinced the Catholic Church to bring Galileo to trial. In 1633 Galileo was convicted of heresy and imprisoned. However he was soon placed under house arrest, and lived comfortably in his villa near Florence. His nun daughter lived with him, and he was able to receive visitors and teach students. He wrote defiant books that were smuggled out to foreign publishers. He died in 1642, the year that Isaac Newton was born.

The Galileo incident is often considered to be part of a continual conflict between science and Christianity, particularly the Roman Catholic Church, but this concept is simplistic. Galileo's book *Dialogue*, published in 1632, infuriated powerful intellectual leaders, including university professors and prominent preaching priests. These intellectuals used their political power to have Galileo's voice suppressed by the government. In Florence, Italy at the time of Galileo, the Catholic Church was the government.

Between 1450 and 1700, thousands of people were executed throughout Europe for witchcraft, in Protestant as well as Catholic areas. When compared with this, Galileo's punishment was not as harsh as it may seem today.

Prior to his astronomy studies with the telescope, Galileo had made other major contributions to science. These included accurate measurements of falling bodies, which proved (in opposition to Aristotle) that bodies of different weight fall at the same rate. Galileo also proved that the path of a projectile follows a parabolic trajectory as it falls to earth.

These measurements of falling bodies by Galileo were later used by Newton to prove that the force of gravity causes a free-falling body to fall with a constant acceleration. This was a key element in the development of Newton's laws of mechanics.

The Discoveries of Isaac Newton

Many people followed Galileo's lead and built telescopes to study the heavens. A telescope of that time had poor resolution because of chromatic aberration. The light from the simple lens that it used does not focus at a single point, because the various wavelengths of light are refracted at different angles. This produces colored fringes around the image.

Modern refracting optical instruments, such as telescopes and cameras, use achromatic lenses, which have two kinds of glass with compensating refraction characteristics. Refraction differences over the visible wavelength band are cancelled, and so chromatic aberration, with its color fringes, is eliminated. The achromatic lens was invented in

1757 by the British optician John Dollond.

Achromatic lenses were not available in the 1600's, and so refracting telescopes at that time were poor. In 1668 Isaac Newton (1642-1727) made a tremendous advance in telescope design by fabricating a parabolic reflecting mirror from metal, and he built from this the first reflecting telescope. All wavelengths are reflected from a mirror at the same angle, and so the primary (or "objective") optical element of his reflecting telescope had no chromatic aberration. The eyepiece used a refracting lens, but chromatic aberration from the eyepiece has much less effect than that from the objective lens. This reflecting telescope was a tremendous improvement, and it made Newton famous.

Newton still wanted to build a refracting lens without chromatic aberration. He performed optical experiments with prisms in order to understand the nature of chromatic effects, so that he could compensate for chromatic aberration. He did not have glass with the proper characteristics to achieve this, and concluded that an achromatic refracting lens could not be built. However in his experiments he made a fundamental discovery that revolutionized the knowledge of light.

A normal light source, such as a candle or the sun, has a broad spectrum covering the visible band. We call this a *white* light, because it looks white to the eye. When a beam of white light is passed through a prism, the wavelengths contained in the white light are refracted at different angles, and thereby form a spectral pattern displaying the colors of a rainbow. It is obvious to us today that a prism separates white light into its component wavelengths, but this concept was not understood at all in Newton's day.

The problem was that the only instrument for detecting optical wavelength effects in Newton's time was the color vision of the eye. Color vision is a very complicated phenomenon, which is not well understood even today. We perceive sound as a succession of tones of different frequencies, but our perception of color is entirely different.

An important aspect of color perception is that vision strongly modifies the optical information that it receives. By means of adaptation processes, the eye compensates to a large extent for the illumination, and so the color of an object appears to be almost independent of the spectrum and intensity of the light that illuminates the object.

Our perception of color has four basic chromatic sensations: red, green, yellow, and blue. It also has the achromatic color sensations, white, gray and black. The four chromatic sensations are arranged in pairs: red-green and yellow-blue. We perceive either red or green, but not a combination of red and green; and we perceive either yellow or blue, but not a combination of yellow and blue.

Therefore we can represent our chromatic sensations in terms of two perpendicular axes, as shown in diagram [a] of Fig 5-1, where green is the negative of red, and blue is the negative of yellow. The chromatic sensation is zero at the origin, and so at that point we perceive the achromatic sensation, which is often gray.

The figure also shows the intermediate chromatic sensations, which are combinations of red, yellow, green, and blue. A combination of red and yellow sensations produces orange; a combination of yellow and green produces yellow-green (olive); a combination of blue and green produces blue-green (aqua); and a combination of blue and red produces violet and purple. This shows that the chromatic color sensations form a circle.

When artists arrange colors into an orderly array, they generally use different spacing in the color circle. A common form of the artist's color circle is shown in diagram [b]. The six colors (red, orange, yellow, green, blue, and violet) are equally spaced around the color circle. This spacing has the advantage that colors opposite one another in the color circle are "contrasting" or "complementary" colors.

When light is projected through a prism, it forms a linear pattern with the colors of a rainbow, as shown in diagram [c]. The chromatic colors are now arranged in the linear sequence: red, orange, yellow, green, blue, and violet. How do we relate this linear pattern of colors projected from a prism to the circular pattern of colors in the artist's color circle?

A reader who has never studied the issue of color perception may feel confused. The sense of color that he has taken for granted suddenly has become an enigma. Yet his confusion is mild in comparison to what was experienced by the associates of Newton in the Royal Society.

Newton had discovered that white light is a combination of different kinds of light of different "refrangibility", which are distinct from one another. However, he did not know that they were of different wavelengths. When he presented his results to the Royal Society in 1672, they were flatly rejected.

The president of the Royal Society, Robert Hooke, had his own theory of color. Hooke believed that light is like a pulse that changes in shape when reflected from a colored object. Hooke was sure that his theory gave a much better answer to why "yellow mixed with blue produces green" than did Newton's explanation.

5. Discovery of the Nature of Light   65

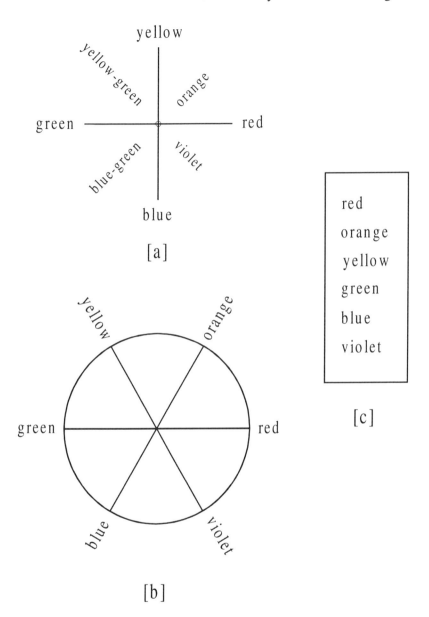

Figure 5-1: Complexity of color perception, which confused Newton's optical research; [a] coordinates of chromatic color perception; [b] the artist color circle; [c] linear rainbow array of colors projected by a prism

As reported by Cohen, Newton published several detailed responses to arguments against his optics research in the *Transactions of the Royal Society*, but was unable to get anyone to even try to repeat his experiments. Finally Newton gave up in disgust, and concluded that: *"A man must either resolve to put out nothing new, or to become a slave to defend it"*.

Newton put his optics research aside to study planetary motions. He and the German mathematician Gottfried Wilhelm Leibniz (1646-1716) invented the mathematical principles of calculus independently, although Newton invented them first. Leibniz refined his calculus to give us an effective mathematical tool. However Newton received the greater fame by applying calculus to develop his laws of mechanics, which he used to calculate planetary orbits.

Newton made strong use of Kepler's three laws in his analysis of planetary orbits, and he also used the measurements of falling bodies made by Galileo. By combining this and related information with calculus, Newton developed his laws of mechanics. He withheld publication until the latest measurements of distances to the sun and moon and the diameter of the earth agreed with his theory. (The distance to the sun measured at that time was within 6 percent of its true value.)

Newton's theory was published in 1687 by the Royal Society, as *Philosophiae Naturalis Principia Mathematica*. Latin was used because scientists in all countries could read Latin. The famous astronomer Edmund Halley (1656-1742) personally paid for this printing.

Newton's theory has two basic laws, which are

(a) **Law of gravitational attraction:** Two bodies are attracted together with a force that is proportional to the product of their masses, and is inversely proportional to the square of the distance between their centers of gravity.

(b) **Law of motion:** The force applied to a body is equal to the mass of the body times its acceleration, where acceleration is the rate-of-change of velocity. *An alternative statement of this is:* the force applied to a body is equal to the rate-of-change of momentum, where momentum is the product of mass times velocity.

Newton also gave the following laws, which are corollaries of his two basic laws:

5. Discovery of the Nature of Light   67

(c) A body at rest, or moving at constant velocity, stays in that condition unless a force is applied to it.

(d) For every action there is an equal and opposite reaction.

By applying calculus to these laws, Newton was able to determine accurately the motions of bodies in the solar system. However when he wrote his *Principia*, which presented his theory of mechanics, he did not use calculus, because scientists at that time did not understand it. Instead he used geometric constructions to achieve the effects of calculus.

Contrary to popular accounts, Newton did not invent the concept of gravitational attraction. Forty-two years before Newton published his *Principia*, Ismaelis Bouillard had postulated that mutual attraction of the planets varies inversely as the square of the distance between them. It is also probable that even before that both Galileo and Kepler understood that gravitational attraction holds the planets in their orbits around the sun, and holds the moon in its orbit around the earth. What Newton presented was a set of precise mathematical laws, with the proof that these laws accurately describe the orbits of the planets around the sun, and the orbit of the moon around the earth.

Newton's *Principia* gave him great fame, and in 1703 he was elected president of the Royal Society. In 1704 Newton published his book *Opticks*, describing his research with prisms. Nearly all of *Opticks* had been written 30 years earlier.

Newton's *Opticks* revolutionized the knowledge of light. Nevertheless, readers of this book in later years have generally misunderstood the meaning it had when it was written. *The basic message of Newton's Opticks was that white light is a combination of different rays having unique refraction characteristics. This finding is so obvious to us today; it is hard for us to realize that it was a revolutionary concept in Newton's time.*

The Wave Nature of Light

Newton was not particularly concerned with the question of whether a light beam was a wave or a stream of particles (or "corpuscles"). However the wave concept seemed dubious. He knew that sound travels through the air as a compression wave by vibrating the air particles, but sound cannot be transmitted where there is no air. What is the mysterious medium, called the *luminiferous aether* that is vibrated by light and allows a light wave to be transmitted over infinite distances in empty space? How can the aether be so thin that it offers no resistance to planets as they travel in their orbits, yet be so stiff that it transmits light vibrations at a fantastically high speed? (The approximate speed of light was known in Newton's time).

Consequently Newton assumed that a light beam was a stream of particles that he called "corpuscles". He postulated that the light rays of different "refrangibility" that make up a beam of white light are corpuscles of different "size". When a beam of white light is passed through a prism, the light corpuscles of different size follow separate paths, and thereby are spread out to form a colored spectrum.

Christiaan Huygens (1629-1695) was a famous Dutch mathematician who lived at the time of Newton. He believed in the wave concept of light, and made extensive mathematical analyses of light-wave propagation.

Because Newton was so famous, his postulate that light is a stream of corpuscles was generally assumed to be correct during the 1700's. However in 1802 Thomas Young (1773-1829) challenged this concept in a Royal Society paper. Young used the theoretical work of Huygens to assist him in his research.

Thomas Young was a physician, but his primary contributions were in other areas. He did studies on the elasticity of materials, and his name is well known today by the term *Young's modulus of elasticity*. He worked on the translation of Egyptian Hieroglyphics from the Rosetta Stone, which carries an inscription written in Hieroglyphics, Greek, and Demotic, which is a simplified script that is related to Hieroglyphics. Deciphering the Rosetta Stone was the key that allowed archaeologists to read the mysterious writings found in Egyptian tombs. Thomas Young achieved the initial translations, and discovered that at least some Hieroglyphics are phonetic symbols. Jean Champollion built on Young's work, and is usually given the credit for deciphering the Rosetta Stone.

However the primary achievement of Thomas Young was to demonstrate the wave nature of light with the experiment shown in Fig. 5-2. Young passed sunlight through a pinhole (A). The light was refracted by (A), and diverged so that it reached pinholes (B) and (C). The light was refracted again at pinholes (B) and (C). The diverging light beams from (B) and (C) interfered with one another, to produce an interference pattern that was projected onto surface (D). (This experiment is often performed with slits, but Young used pinholes.)

This optical interference pattern proved that light is a wave. With this experiment Young was able to measure the wavelength of light.

In 1802 to 1804 Young published the results of his light-interference experiments in the *Transactions of the Royal Society*. The response was a merciless attack against him. This included two anonymous discussions of Young's work in the *Edinburgh Review*, probably written by Lord Brougham, which ridiculed Young for being so arrogant as to question the great Sir Isaac Newton. This included the following:

> *"We wish to raise one feeble voice against innovations that can have no other effect than to check the progress of science, and renew all those wild phantoms of the imagination which Bacon and Newton put in flight from her temple. The paper contains nothing that deserves the name of either experiment or discovery."*

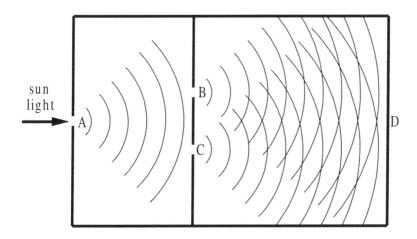

Figure 5-2: Light interference experiment by Thomas Young

A few years later a young French physicist, Augustin Fresnel (1788-1827), who was unaware of Young's work, began to perform similar light interference experiments. His friend, Dominique Arago, brought him in contact with Thomas Young. In 1819 Fresnel won the Scientific Competition Award by the French *Academie des Sciences* for his research on light-wave interference. This involved a precise experiment combined with sophisticated mathematical analysis. This award placed the wave theory of light on a solid footing, but opposition to the theory continued for years.

Augustin Fresnel died in 1827 at age 39, and the world lost a great scientist. He had been sickly for years with tuberculosis.

Newton has often been criticized for insisting on the corpuscular theory of light, but this criticism is unjustified. When Newton described his optical experiments, he did not particularly care whether light was a wave of a stream of particles. His "rays of different refrangibility" could just as well be *waves of different wavelength or particles of different size*. He needed a theoretical model to explain his experiments, and he picked the particle ("corpuscle") concept because it seemed more believable.

Because of the stubbornness of Robert Hooke, president of the Royal Society when Newton did his optical research, and the sheep-like mentality of others in the society, Newton's optical findings were rejected. Nearly all of Newton's *Opticks* was written at that time, but it was not published until 30 years later. After publication, many people immediately repeated and extended Newton's experiments, and the science of optics advanced rapidly.

Newton's revolutionary idea was that white light is a combination of unique rays having different refraction characteristics. After a few years this idea became absurdly obvious, and so readers of Newton's *Opticks* gradually missed its real point. Instead many came to assume incorrectly that this document was written as a treatise expounding the corpuscular theory of light.

Because of Newton's enormous prestige, his casual corpuscular postulate became enshrined as absolute truth. Thomas Young faced strong opposition to his demonstration of the wave nature of light because it conflicted with the preconceived beliefs of many of his contemporaries. Young's research did not conflict in a substantive way with the statements made by Newton in his book *Opticks*.

Thus Thomas Young and Augustin Fresnel proved that light is a wave. But how does it propagate? The mysterious aether that transmits light waves was as enigmatic as ever.

Electromagnetic Wave Concept

During the 1800's great advances were made in our knowledge of electricity and magnetism, from research performed by Coulomb, Volta, Oersted, Gauss, Weber, Ampere, Faraday, and others. These electrical and magnetic findings were tied together in 1873 by James Clerk Maxwell (1831-1879), and presented as a set of *electro-magnetic* equations. These equations deal with electrical and magnetic "fields", and so are called *electromagnetic field equations*.

Maxwell predicted from his equations that an electromagnetic

5. Discovery of the Nature of Light 71

"radio" wave can be generated, consisting of oscillating electric and magnetic fields. The predicted speed of this wave (based on measured electrical and magnetic parameters) was the same as the measured speed of light. Therefore Maxwell concluded that light must be an electromagnetic wave of very high frequency.

In 1888 Heinrich Hertz (1857-1894) generated the electromagnetic "radio" wave that Maxwell had predicted. However it was not until 1895 that Guglielmo Marconi (1874-1937) achieved practical communication with radio over a few kilometers. Marconi used a radio wave of much lower frequency than Hertz. Marconi performed his first radio experiments in 1890 when he was only 16. He established telegraphic radio communication links across the English Channel in 1899, from England to St. John's, Newfoundland in 1901, and from England to Cape Cod, Massachusetts in 1903.

Maxwell's electromagnetic field equations are used extensively today in complicated analyses. They are the foundation that allows engineers to design many electrical devices, such as the antennas for transmitting and receiving radio, television, and radar signals. Marconi must have studied Maxwell's equations intensely to develop his very sophisticated and large antenna designs, which were the first antennas for transmitting and receiving radio signals.

The following discussion gives a simple physical explanation of an electromagnetic wave. In diagram [a] of Fig. 5-3, an electrical current is sent through a coil of wire to generate a magnetic field. The dashed lines show the lines of magnetic flux that make up the field. One can visualize this magnetic field by sprinkling iron filings in the region of the coil, and watching the filings line up with the lines of magnetic flux.

If a compass is placed in the magnetic field as indicated in the figure, its north-seeking arrow will point in the direction that is shown. If the electric current in the coil is reversed, the compass will point in the opposite direction. This indicates that the magnetic field has an absolute direction.

Diagram [b] of Fig. 5-3 shows a transformer. Two coils of wire are wound around a core of iron. A generator produces an oscillating electrical current that is fed into coil (1), which has N_1 turns of wire. The current produces an oscillating magnetic field inside the iron core. The oscillating magnetic field generated by coil (1) is conducted by the iron core to produce an oscillating magnetic field within coil (2), which has N_2 turns of wire. The oscillating magnetic field within coil (2) generates an oscillating voltage across the terminals of coil (2), which causes an oscillating electric current to flow into the load.

The voltage generated across the terminals of coil (2) is equal to the voltage applied to coil (1) multiplied by the ratio N_2/N_1 of the turns in the two coils. Transformers are employed to change the voltage of alternating electricity. For example, electrical power is transmitted over long distances at high voltage to reduce electrical current and thereby minimize power loss in the wires. Near each house is a transformer that converts the high transmission voltage to the much safer 120 volts that is used in our homes.

The voltage applied to coil (1) of the transformer can be considered to be an electric field. This oscillating electric field generates the oscillating magnetic field that is conducted within the iron, and the oscillating magnetic field generates the oscillating electric field across coil (2). *Thus an oscillating electric field produces an oscillating magnetic field, and an oscillating magnetic field produces an oscillating electric field.*

This principle is the basis for an electromagnetic light wave or radio wave. An electromagnetic wave consists of electric and magnetic fields that oscillate at right angles to one another. The oscillating electric field generates the oscillating magnetic field, and the oscillating magnetic field generates the oscillating electric field. The two fields, oscillating at right angles, support one another, and together they form a packet of energy that propagates in a direction that is perpendicular to the oscillation plane of the two fields.

This concept is illustrated in Fig. 5-4. The electric field E vibrates up and down in the direction of the x-axis, which we assume to be vertical. The magnetic field H vibrates back and forth in the direction of the y-axis, which we assume to be horizontal. The solid E arrow shows the direction of the electric field during one half cycle, and the dashed E arrow shows the direction in the next half cycle, when it is pointing in the opposite direction. Similarly, the solid and dashed H arrows show the directions of the magnetic field in alternate half cycles. The oscillating electric and magnetic fields are out of phase with one another, so that when one field is a maximum the other is zero. The oscillating electric and magnetic fields support one another, to form an electromagnetic wave that propagates in the horizontal z-direction. The x-y plane (within which the electric and magnetic fields oscillate) is perpendicular to the direction z in which the wave propagates.

Light is a packet of electric and magnetic fields that travels like a particle at the speed of light. Light also acts like a wave, because the electric and magnetic fields oscillate as they propagate. However, light is not like a sound wave or a wave on water, which propagate by vibrating

5. Discovery of the Nature of Light   73

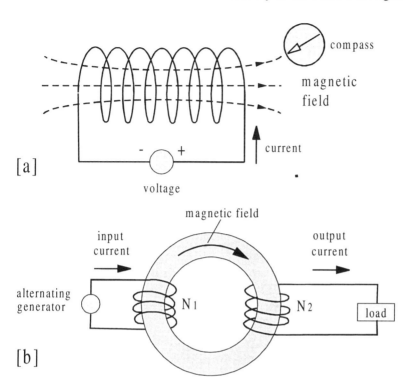

Figure 5-3: Electromagnetic circuits; [a] magnetic field generated by magnetic coil; [b] transformer circuit.

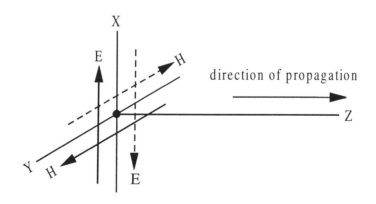

Figure 5-4: Oscillating electric and magnetic fields that form an electromagnetic wave.

a medium. Light can travel through a vacuum, but sound cannot because there is no medium in the vacuum to vibrate.

A light wave propagates by means of oscillating electric and magnetic fields, not by vibrating a mysterious substance called the *luminiferous aether*. Maxwell's electromagnetic field equations eliminated the need for an aether in light propagation, thereby removing the stumbling block that had caused Newton to reject the concept that light is a wave. Nevertheless Maxwell continued to assume in his writings that the aether still existed.

Search for the Velocity of the Luminiferous Aether

Why did Maxwell maintain the aether concept, even though the aether made no sense physically, and was not needed in the propagation of an electromagnetic wave? The following is a likely explanation.

A sound wave travels with respect to the air, which vibrates, and so the velocity of the sound wave is measured relative to the air. Remember that sound cannot propagate where there is no air.

What happens with a light wave? Suppose that the light transmitter and light receiver are moving relative to one another at a velocity that is appreciable relative to the speed of light. To which reference is the speed of light measured? Is the light speed measured relative to the transmitter; or is it measured relative to the receiver; or (like sound) is it measured relative to the aether medium through which the light is propagating?

If we eliminate the aether medium, how do we measure the speed of light when there is a velocity between the light transmitter and the light receiver? Do we measure the speed of light relative to the body that transmits the light, or do we measure it relative to the body that receives the light? Without an aether there seemed to be no rational way to specify the speed of light.

This enigma is probably what led Maxwell to include the aether concept in his writings. This hypothesis is supported by the fact that Maxwell's influence was a major factor leading Michelson and Morley to perform the famous Michelson-Morley experiment for measuring the velocity of the aether.

The earth rotates around the sun at a velocity of 30 km/sec, which is 0.01 percent of the speed of light (300,000 km/sec). Let us assume initially that the aether is moving with the sun. If light is travelling at a constant velocity relative to the aether, the velocity of light measured on the earth from a given star should vary over a 12-month period by ±0.01 percent. Even if the aether is not stationary relative to the sun, the aether velocity should still be measurable by comparing data taken month by month along the earth's orbit.

An instrument was made by Michelson and Morley using wave interference techniques that could measure the difference in the speed of light in perpendicular directions with accuracy much greater that 0.01 percent. The experiment was performed in 1887, and the results were completely negative. There was no detectable variation in the speed of light from any direction, at any point in the earth's orbit.

Many hypotheses were proposed to explain the negative findings of the Michelson-Morley experiment. Finally George FitzGerald and Hendrik Lorentz (1853-1928) independently postulated that a physical object contracts when it moves in the direction of the aether. In 1904 Lorentz generalized his postulate by presenting a set of formulas called the *Lorentz transformation* equations, which relate measurements made by observers moving at different velocities relative to the aether. Lorentz showed that Maxwell's electromagnetic field equations are unaltered when transformed by his formulas.

Nevertheless neither Lorentz nor FitzGerald truly understood the nature of the problem. This answer was provided in 1905 by the famous paper on relativity published by Albert Einstein (1879-1955). In that same year Einstein also published highly significant papers on the photoelectric effect and on statistical mechanics. He was awarded the Nobel Prize for his paper on the photoelectric effect. Einstein did not receive a Nobel Prize for his much more important research on relativity, because it was too controversial at the time.

Einstein recognized that an aether medium that provides an absolute velocity reference for light makes no sense. He ignored the Michelson-Morley experiment, and based his reasoning on fundamental principles. He concluded that the transmitter and receiver of a light wave must measure the same value for the speed of light, regardless of the velocity between the two. There is no such thing as absolute velocity; there is only relative velocity. *We can specify the relative velocity between two observers, but not the absolute velocity of either one.*

How can two observers that are moving relative to one another measure exactly the same value for the speed of light? To answer this question, we must consider how clocks and measuring rods are used to perform the measurements, and how the clocks are synchronized. These issues, which Einstein discussed in his 1905 paper on relativity, will be explained in a simple manner in Chapter 6.

The relativity concepts that Einstein presented in his 1905 paper were concerned with observations made from two coordinate systems moving at constant velocity relative to one another. Later in 1916 Einstein generalized his relativity concepts to include coordinate systems that are accelerating. The earlier relativity theory became known as *special relativity*, and the generalization to include acceleration was called *general relativity*.

Einstein showed that the effects of gravity and acceleration are indistinguishable. Hence his general theory of relativity, dealing with accelerating coordinate systems, is fundamentally a theory of gravity. General relativity theory requires complicated tensor mathematics, which uses the Riemannian geometry of curved space, and so is very much more complicated than special relativity.

Was the Wave Theory of Light Correct?

Einstein received the Nobel Prize for his 1905 paper on the photoelectric effect, which proved that light is quantized into elements called photons. He showed that light acts like a stream of corpuscles (or photons), as well as like a wave. *Therefore we now know that the corpuscular theory of light and the wave theory are both correct.*

On the other hand, a light wave does not propagate by vibrating a medium, as does a mechanical wave, like a wave on water or a sound wave. As explained earlier, when a wave travels over the surface of a pond, the individual water molecules oscillate in small elliptical paths, and consequently move very little. It is the mechanical energy of the wave, not the water, that actually travels across the pond.

In contrast, a light source emits electric and magnetic fields that propagate like corpuscles at the speed of light. Light has the properties of a wave because these travelling electromagnetic fields oscillate as they propagate. However the traditional wave theory of light envisioned light to be the vibration of a medium. Even though light acts like a wave, it does not propagate by vibrating a medium, and so the traditional wave theory of light was wrong.

Belief in the physically impossible *luminiferous aether* was based on centuries of tradition that regarded all waves to be vibrations of a medium. This belief remained strong long after Maxwell had discovered the principle of an electromagnetic wave.

Chapter 6

Einstein Special Theory of Relativity

To investigate cosmology effectively, we must understand the Einstein theory of relativity. As we will see, this is not as formidable as one might assume. The reader can achieve a physical understanding of relativity from simple discussions. We start by considering the basic version of relativity, which was presented in 1905. After Einstein developed his *General Theory of Relativity* in 1916, his basic theory became known as *Special Relativity*.

Relativity theory evolved from an enigma concerning the speed of light, and so we start our investigation by examining the process of measuring the speed of light. To understand this we first see how the speed of sound is measured.

Measuring the Speed of Sound

Sound travels as a compression wave through the air. The speed of sound varies with the temperature of the air, and with altitude, which affects the pressure. The speed of sound at sea level is approximately *330 meters per second*. To simplify our calculations, we round this off to *300 meters per second*.

We can determine the speed of sound by measuring the time for a sound wave to travel the length of a 3-meter measuring rod, as shown in Fig 6-1a. This time is equal to the length of the rod (3 meters) divided by the speed of sound (300 meters/sec), which gives 1/100 second. It is convenient to express this time in terms of milliseconds, where one millisecond is one-thousandth of a second (0.001 sec). It takes 10 milliseconds (0.010 second) for the sound wave to travel the 3-meter length of the measuring rod.

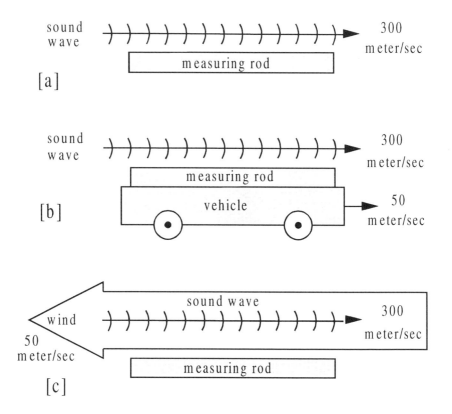

Figure 6-1: Measuring the speed of sound; [a] measuring rod fixed, no wind; [b] measuring rod moving with sound at 50 meter/sec; [c] measuring rod fixed, with wind at 50 meter/sec blowing opposite to sound propagation.

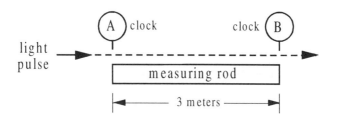

Figure 6-2: Measuring the speed of light

Let us mount the measuring rod on a vehicle moving away from the sound at 50 meters per second (112 mph), as shown in diagram b. The relative speed between the sound wave and the measuring rod is now 250 meter/sec. Dividing the 3-meter rod length by this 250 meter/sec speed gives a time interval of 0.012 seconds (12 milliseconds) for the sound wave to travel the length of the measuring rod.

Finally, let us assume that wind is blowing at 50 meter/sec against the sound wave, as shown in diagram c. The measuring rod is fixed on the ground. The relative velocity between the sound wave and the measuring rod is now 250 meter/sec, just as in case b. Therefore in case c it takes 12 milliseconds for the sound wave to travel the length of the measuring rod.

Sound travels with respect to the air. Consequently the time for the sound to travel the length of a measuring rod depends on the relative velocity between the rod and the air. These examples may seem trivial, but they help to explain the much more complicated effects that occur when we measure the speed of light.

Measuring the Speed of Light

Light travels about one million times faster than sound. Its speed is 300 million meters per second. This can be expressed as 300 meters per microsecond, where one microsecond is one-millionth of a second. It takes 1/100 microsecond for light to travel the length of a 3-meter measuring rod. It is convenient to express this time in terms of nanoseconds, where one nanosecond is one-thousandth of a microsecond. It takes 0.010 microsecond, or 10 nanoseconds, for light to travel the length of a 3-meter measuring rod.

Because light travels so fast, we must be careful in measuring the time for the light to travel the length of the measuring rod. This can be achieved as shown in Fig. 6-2. Clocks 1 and 2 are placed at the leading and trailing ends of the measuring rod. The two clocks run at exactly the same rate, and the clocks are accurately synchronized. Synchronization could be achieved by locating the two clocks at the same point, setting their readings equal, and then moving them to the ends of the measuring rod. When the light reaches the leading edge of the measuring rod, clock 1 is read. When the light reaches the trailing edge, clock 2 is read. The two clock readings are subtracted to obtain the time for the light pulse to travel the length of the measuring rod. This time difference between the two clocks is 10 nanoseconds.

80 *How Was Our Universe Created?*

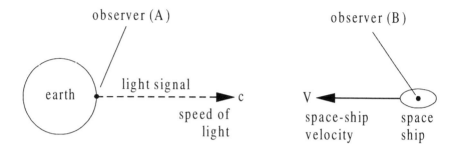

Figure 6-3: Measuring the speed of a light signal transmitted from earth to a space ship, which is returning at a velocity V equal to 60 percent of the speed of light c.

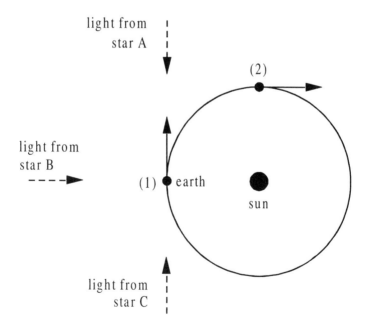

Figure 6-4: Geometry associated with Michelson-Morley experiment

6. Einstein Special Theory of Relativity   81

Now let us consider a fictitious space travel episode. As shown in Fig 6-3, a space ship is returning from an interstellar voyage, and is now traveling toward earth at 60 percent of the speed of light. Since the speed of light is 300 meters per microsecond, the space ship velocity is 180 meters per microsecond. A light pulse is transmitted from earth, which leaves earth at a speed of 300 meters per microsecond. The earth observer A measures the speed of this light pulse in the manner that we have described. The observer B on the space ship measures the speed of the light that he receives from earth. What speed of light does the space ship observer measure?

The answer is, "300 meters per microsecond", exactly the same speed of light measured by the earth observer. Both observers have identical equipment, and both find that it takes exactly 10 nanoseconds for the light to travel the length of a 3-meter measuring rod.

We do not have a space ship that travels at 60 percent of the speed of light to try this experiment. However, we can perform a similar experiment by measuring the light received on earth from stars in different directions. Our earth is our own space ship. As the earth rotates around the sun, it travels at 30 km/sec, which is 0.01 percent of the speed of light. This velocity is much less than 60 percent of the speed of light, but is adequate to prove our point.

In the 1800's it was generally believed that light travels through space relative to a mysterious medium called the aether, just as sound travels relative to the air. Figure 6-4 shows the earth rotating around the sun with velocity V. Let us assume temporarily that the aether is stationary relative to the sun. If light moves relative to the aether with velocity c, the speed of light measured on earth from star B when the earth is in position (1) should be c. The speed of light measured from star A should be equal to the sum (c + V), and the speed of light measured from star C should be (c - V). If the aether is not stationary relative to the sun, one should be able to determine the velocity of the aether wind by measuring the speed of light from different stars at various points in the earth's orbit around the sun.

This was the concept behind the famous Michelson-Morley Experiment, which was performed in 1887. Wave interference techniques were applied, which could measure differences in the speed of light to an accuracy far better than 0.01 percent, which is the velocity of the earth relative to the speed of light. The experiment was implemented, and the results were completely negative. There was no detectable variation of the speed of light, for light received from any star at any point in the earth's orbit.

The Einstein Theory of Relativity

Many unsatisfactory postulates were proposed to explain the negative results of the Michelson-Morley experiment. Finally, George Fitzgerald and Hendrik Lorentz independently proposed that the negative results can be explained by postulating that mechanical dimensions contract when a device moves with respect to the aether. In 1904, Lorentz generalized his postulate to develop a set of equations, called the Lorentz transformations, which show how spatial and time measurements must change when one moves relative to the aether. Lorentz concluded that it was probably impossible to perform an experiment that could detect motion relative to the aether.

In his 1905 paper on relativity, Einstein made sense out of this confusion. Lorentz quickly accepted Einstein's answer, even though it conflicted with his own thinking.

Einstein concluded that the aether does not exist, and that all observers moving at constant velocity measure exactly the same value for the speed of light. There is no such thing as absolute velocity; there is only relative velocity. When there is relative velocity between two observers, their measurements of distance and time appear to be different. Nevertheless, apparent affects are real effects. Reality is different for the two observers.

Einstein applied the same Lorentz transformation equations derived a year earlier by Lorentz, but he interpreted the results in a radically different manner. With his approach, Einstein was able to derive physical predictions that were not obtained by Lorentz.

Going back to Fig. 6-3, let us apply the Einstein relativity theory to explain why the earth observer A and the observer B on the space ship measure the same value for the speed of light. Relative to the earth observer, the clocks on the space ship appear to run slow by a factor K equal to $\sqrt{[1 - (V/c)^2]}$, where V is the relative velocity between the earth and the space ship, and c is the speed of light. Since the velocity V is 60 percent of the speed of light c, V/c is 0.6 and this factor K is equal to

$$K = \sqrt{[1 - (V/c)^2]} = \sqrt{[1 - (0.6)^2]} = \sqrt{[0.64]} = 0.8$$

To the earth observer, the clocks on the space ship appear to run at 80 percent of the normal rate. Also the measuring rod appears to contract to 80 percent of its normal length when held in the direction of relative velocity. To the earth observer, the 3-meter measuring rod on the space ship appears to be 2.4 meters long, which is 80 percent of 3 meters. The third and final effect is that the two clocks on the space ship, which observer B considers to be synchronized, are not synchronized relative to the observer A on earth. The synchronization error in these clocks seen by observer A is

Synchronization error = $(V/c)\Delta t = 0.6\, \Delta t$

The quantity Δt is the time for light to travel between the two clocks, as seen by observer B. We use the Greek letter Δ (delta) to denote a difference, and so Δt represents a difference in time t. Observer B finds that it takes 10 nanoseconds for light to travel the length of his measuring rod, and so Δt is 10 nanoseconds. Hence observer A sees a synchronization error between the two clocks that is 60 percent of 10 nanoseconds, which is 6 nanoseconds. Nevertheless, Observer B considers his two clocks to be exactly synchronized.

To observer A on earth, the light pulse should travel relative to the space ship at a velocity equal to (c + V), which is (300 + 180), or 480 meters per microsecond. To observer A, the space-ship measuring rod appears to be 2.4 meters long, and so the time for the light pulse to travel the length of the space-ship measuring rod is equal to the rod length (2.4 meters) divided by the speed of light relative to the space ship (480 meters per microsecond). Hence it should take 2.4/480 microsecond for the light pulse to travel the length of this measuring rod. This is 1/200 microsecond, or 5 nanoseconds. This shows that to observer A the light pulse appears to take 5 nanoseconds to travel the length of the measuring rod on the space ship.

However, the clocks on the space ship appear to run at 80 percent of the normal rate. Consequently, observer B should interpret this 5-nanosecond time interval to be 4 nanoseconds.

To the earth observer A, the two clocks used by B are out of synch by 6 nanoseconds. Adding this 6-nanosecond synchronization error to the 4-nanosecond time interval gives 10 nanoseconds. Hence, from the point of view of earth observer A, observer B mistakenly thinks that it takes (6 + 4) nanoseconds, or 10 nanoseconds, for the light to travel the length of the measuring rod. In this manner we can explain why both observers find that it takes 10 nanoseconds for light to travel the length of the measuring rod.

There is no such thing as absolute velocity; there is only relative velocity. The earth observer A can assume that the earth is stationary and the space ship is moving at 60 percent of the speed of light.

Likewise, the space ship observer B can assume that the space ship is stationary, and the earth is moving at 60 percent of the speed of light. If the space ship sends a light pulse to earth, the preceding discussion can be applied to the space ship rather than to the earth, by considering the space ship observer to be A and the earth observer to be B.

Implications of Relativistic Effects

When we consider how effects appear to the earth observer and the space ship observer, we should realize that these apparent effects are real. All reality is relative. There are no absolutes. The measuring rod on the space ship is 3 meters long as far as the space ship observer is concerned, and it is 2.4 meters long as far as the earth observer is concerned. Both statements are correct. This is the fundamental principle of relativity.

These confusing concepts may be difficult to accept, but we have no choice. Without these concepts we are unable to explain why both observers measure exactly the same value for the speed of light.

The fact that two clocks that are synchronous relative to one observer are not synchronous relative to another observer tells us that there is no such thing as absolute time. Events that are simultaneous to one observer are generally not simultaneous to another observer moving at a different velocity.

This indicates that measurements in time and space cannot be considered separately. In relativity, time and spatial measurements are combined together into a four-dimensional space-time specification. However, this does not mean that we should regard time to be a mysterious fourth dimension that is equivalent to a spatial dimension. It merely means that time and space must be considered together to obtain a precise specification. A time interval between two events experienced by one observer can appear to be a distance interval to another observer.

Equivalence Between Energy and Matter

These relativity concepts were used by Einstein to derive a number of physical principles. One of these is that matter of mass M is equivalent to an energy E given by

$$E = Mc^2$$

This famous Einstein formula showed that mass can be converted into energy. It explains the source of the energy radiated by the sun, and eventually led to the atomic and hydrogen bombs. The equation shows that one gram of mass is equivalent to 25 million kilowatt hours of energy. This means that 25 million kilowatt hours of energy are released for every gram of matter that is converted into energy.

The sun achieves its energy by fusing four hydrogen atoms to form one helium atom. The atomic weight of helium is 3.971 times that of hydrogen, and so the ratio of helium mass to hydrogen mass is 3.971/4, which is 0.99275. Thus the helium atom has 99.275 percent of the mass of the four hydrogen atoms that form it. The remaining 0.725 percent of the hydrogen mass is converted into energy. Taking 0.725 percent of 25 million kilowatt hours gives 181,250 kilowatt hours. This shows that the conversion of one gram of hydrogen into helium releases 181,250 kilowatt hours of energy. One gram is 1/3 of the weight of a United States penny.

It is remarkable that the geometric concepts of special relativity allowed Einstein to explain the source of the enormous energy radiated by our sun. *The fact that matter can be converted into energy in accordance with this Einstein formula, tells us that our relativity principles are correct. It demonstrates that all Reality is Relative.*

The Principle of Covariance

The relativity principles that we have applied to relate the measurements of the observer in the space ship to the observer on earth are expressed in rigorous mathematical form by a set of formulas known as the Lorentz transformation equations. These equations were first derived in 1904 by Hendrik Lorentz (1853-1928), and so bear his name. However Lorentz interpreted his equations in terms of motion relative to the hypothetical aether. It was Einstein who established the basic principles of relativity.

Let us consider the application of the Lorentz transformation equations to physical problems. We can imagine that each observer in our discussion is replaced by a set of coordinates. Instead of considering measurements made by particular observers, we can consider measurements made relative to coordinate systems at the locations of the observers. The Lorentz transformation equations allow one to translate measurements made relative to one coordinate system to another coordinate system, in such a manner that physical consistency is maintained.

We have been investigating the resolution of an inconsistency in measuring the speed of light. However, the Lorentz transformation equations have more general applicability. Maxwell's electromagnetic field equations characterize the physical laws of electricity and magnetism. Lorentz proved that Maxwell's equations are invariant when modified by his Lorentz transformation equations. What this means is that the two observers, moving at different velocities, experience exactly the same electrical and magnetic laws when they make measurements relative to their separate coordinate systems.

Thus the Lorentz transformation equations allow one to translate physical measurements from one coordinate system to another in such a manner that the same physical laws are experienced by observers in the two coordinate systems.

This concept can be stated in a general manner as the *Principle of Covariance*. This principle states that the laws of physics should be expressed in such a manner that that they are independent of the coordinate system. The fundamental goal of Einstein's research in developing his general theory of relativity was to achieve *Covariance* in his transformation equations.

Special relativity deals with measurements made by observers in separate coordinate systems moving at different but constant velocities. Under this condition, the speed of light is the same in both coordinates. However when the velocity changes, so that acceleration occurs, special relativity no longer applies exactly, and the speed of light can be different in the two coordinate systems. Einstein showed that acceleration and gravity are equivalent. Therefore the Lorentz transformation equations of special relativity do not apply exactly in the presence of a gravitational field.

Although the Lorentz transformation equations of special relativity do not apply exactly in a gravitational field or under acceleration, they still give extremely close approximations for almost all applications within our solar system. It is only in very special cases that these theoretical limitations are important.

Nevertheless the theoretical limitations of special relativity are fundamental, and Einstein felt compelled to eliminate them. In 1916, after 11 years of hard work, he achieved this with his general theory of relativity. His general theory achieves *Covariance* of physical laws when the coordinate systems are accelerating or are in gravitational fields.

Chapter 7

General Relativity

Generalizing the Relativity Principle

Einstein derived several profound conclusions from his basic relativity concept, but he soon found that it had a serious weakness. The theory applies exactly only when the velocities of the two observers are constant. When acceleration occurs, which means that the velocity is changing, Einstein found that the speed of light is not exactly constant. He also concluded that acceleration and gravity are equivalent, and so a gravitational field causes the speed of light to change. Since constancy of the speed of light is the unifying principle for his basic theory of relativity, Einstein needed a new principle to generalize the relativity concept. He found this in the complex mathematics of *tensor analysis*.

Tensor analysis evolved from the *Riemannian geometry* of curved space, which was first presented in 1852 by the German mathematician Bernhard Riemann (1826-1866). Riemann specified curved space by means of the *geodesic* and the *metric equation*, which describe the shortest distance between two points in curved space. Riemann contracted tuberculosis in 1862, and died four years later at age 39.

The Italian mathematician Gregorio Ricci (1853-1925) used the principles of Riemann as the foundation to develop a comprehensive mathematical theory called the *absolute differential calculus*. This theory, now known as *tensor analysis*, was published in 1901.

Tensor analysis has a formula for translating a tensor from one coordinate system to another. Therefore, by expressing relativity principles in terms of tensors, Einstein had the basis for achieving *covariance*, whereby the same laws of physics hold in all coordinate systems, regardless of velocity and acceleration, and regardless of gravitational fields. However, tensor analysis is very complicated. It took

11 years of intensive research before Einstein was able to publish his *General Theory of Relativity* in 1916. His basic relativity theory presented in 1905 was then called *Special Relativity*.

Einstein concluded that the concept of gravitational force presented by Newton was incompatible with relativity, because it represents a force operating instantaneously at a distance; whereas relativistic effects propagate at the speed of light. Instead, Einstein described gravity in terms of a curvature of space. Matter causes space to be curved, and this space curvature produces the effect that we interpret as gravitational force. The Riemannian principle for specifying curved space, which is incorporated in tensor analysis, provided the basis for characterizing gravity in general relativity.

In 1916 Einstein presented his general theory of relativity, which was specified in terms of a tensor formula called the *Einstein gravitational field equation*. This tensor formula represents 10 independent equations. Because of the great complexity of this formula, Einstein was only able to derive approximate solutions from it. However, Karl Schwartzschild (1873-1916), who was cooperating with Einstein, applied the Einstein formula to a simple physical model of a star, and thereby achieved an exact solution. Einstein also published the Schwartzschild solution in 1916. Sadly, Karl Schwartzschild died suddenly from disease even before his famous analysis was printed. He was a German army officer at the Russian front during World War I.

General relativity theory incorporates the special relativity effects caused by velocity and the general relativity effects caused by gravity and acceleration. Within the weak gravitational fields of our solar system, the relativistic effects due to gravity are very small, but are measurable. Based on the Schwartzschild solution, Einstein devised the following three tests to verify his general theory of relativity:

(1) When a light ray passes close to the sun, it should be deflected by 1.8 arc seconds.

(2) A gravitational field causes a clock to run slower, and therefore causes the excited elements on the surface of the sun to oscillate at lower frequency, thereby generating spectra of longer wavelength. Consequently the spectral lines are shifted toward the red end of the spectrum. This gravitational redshift of radiation from our sun should be 2.1 parts per million (i.e., 2.1×10^{-6}).

(3) The planet Mercury has a highly elliptical orbit. The axis of the

Mercury orbit advances (or rotates) by 1.39 arc seconds per orbit. Of this rotation of the orbit axis, 92.5 percent can be explained with Newton's laws by considering the gravitational attraction of other planets. A residual error of 0.10 arc second per orbit remained, which was explained by the Einstein general theory of relativity.

These three tests were implemented, and the results established the validity of the Einstein general relativity theory.

These measurable effects of general relativity are very small: an advance of only 0.10 arc second per orbit of Mercury; a 1.8 arc second deflection of a light beam passing close to the sun, and a gravitational redshift of only 2.1 parts per million in light emitted from the sun. Hence one might wonder why Einstein worked so hard to achieve his theory, and why general relativity is so highly regarded. *The answer is that this generalization was essential to provide a solid theoretical foundation for the relativity principle that is embodied in special relativity.*

When the predictions of general relativity were verified, Einstein achieved great fame. After that time, Einstein did little with his general relativity theory. Special relativity is very much easier to apply, and has wide application. During Einstein's lifetime, general relativity, with its very complicated tensor analysis, served primarily as a theoretical foundation for justifying special relativity.

Einstein could apply general relativity only to very simple cases. With more complicated applications, the equations of general relativity can yield millions of terms, and so could not be solved analytically. In the 1960's, a decade after Einstein's death, powerful computers became available, which could apply the Einstein general relativity theory to complicated physical models. Since then, hundreds of mathematicians, physicists, and engineers in academic positions have devoted their careers to computer solutions of the Einstein gravitational field equation. This effort has been devoted to Big Bang studies of cosmology, because that is the only area that can use this expertise.

Applying the Equivalence Principle

Einstein concluded that the effects produced by gravity and acceleration are equivalent. He examined the principles of special relativity under conditions of acceleration, and related these results to a gravitational field in order to generalize his relativity theory. Let us see how he did this.

90 How Was Our Universe Created?

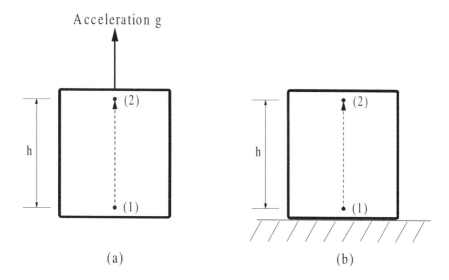

Figure 7-1: Elevator (a) is accelerating in free space; elevator (b) is fixed on earth. Light travels from point (1) to point (2).

As shown in Fig. 7-1, Einstein considered two identical elevators. One rests on the ground, as shown in (b), and the other, shown in (a), is located in space, where gravitational force is negligible. The elevator in space is being pulled upward with the same acceleration of gravity "g" experienced on earth. When Einstein's elevator is considered today, it is generally assumed that a rocket motor under the elevator (a) is pushing the elevator upward with the acceleration g.

Let us consider what we mean by the acceleration of gravity "g" experienced on earth. When one allows an object to fall on earth, it accelerates downward with the acceleration of gravity, which is denoted g. The value for g is approximately 32 ft/sec per second, or 10 meters/sec per second. This means that after one second the velocity of the object is 10 meters/sec, and after 2 seconds it is 20 meters/sec. The velocity of a falling object is proportional to time. This assumes that aerodynamic forces exerted by air on the object are negligible.

When an object is not allowed to fall, the object exerts a force on the floor under the object, which we call weight. The weight W of an object is equal to the mass M of the object multiplied by the acceleration of gravity g.

In a space vehicle orbiting around the earth, the astronauts experience weightlessness. They can float around the cabin. They still have mass M, but they experience no weight W because the effective acceleration of gravity is zero. The earth is still exerting gravitational pull at that point, but the space vehicle is orbiting around the earth so rapidly that centrifugal force cancels the force of gravity.

The elevator in diagram (a) of Fig 7-1 is being pushed upward with a rocket at the same acceleration of gravity g that is experienced on earth. An object within this elevator would exert the same weight force on the floor of the elevator (a) that it would exert if it were located within the fixed elevator (b) located on earth. The conditions within elevators (a) and (b) are the same. If the elevators are closed, a scientist could not tell from experiments performed within the two elevators whether he is in elevator (a) or elevator (b). This is the basis for the *Principle of Equivalence* proposed by Einstein. This principle allowed Einstein to relate the effects of acceleration to those produced by gravity.

Suppose a light pulse is emitted from the floor of elevator (a) at point (1) and is received at the ceiling of the elevator at point (2). The height of point (2) above point (1) is denoted h. The time for the light pulse to travel from point (1) to point (2) is approximately equal to h/c, where c is the speed of light. During the propagation time of the light pulse, the upward velocity of the elevator will have increased because the elevator is accelerating. The increase of velocity is denoted ΔV. This velocity difference ΔV is equal to the acceleration of gravity (g) multiplied by the time interval (h/c) for light to travel from point (1) to point (2). Hence ΔV is equal to (gh/c). The velocity ratio $\Delta V/c$ is equal to (gh/c^2).

Because the velocity changes during the propagation time of the light pulse, there is an effective velocity difference between point (1) and point (2). This velocity difference causes a Doppler wavelength shift of the light. We use the Greek letter λ (called lambda) to represent wavelength. (Note that the Greek letter λ is equivalent to the Roman letter l, and so is a good symbol for "wave-Length".) The emitted wavelength is denoted λ, and the shift in wavelength is denoted $\Delta\lambda$. The ratio $\Delta\lambda/\lambda$ is approximately equal to $\Delta V/c$, which is the velocity change ΔV divided by the speed of light c. Therefore, the wavelength ratio $\Delta\lambda/\lambda$ is approximately equal to (gh/c^2).

Relative to a light pulse traveling from point (1) to point (2), the points appear to be moving apart. Consequently, the Doppler effect shifts the light spectrum toward longer wavelength. The spectrum is shifted toward the red, and so we call this a "redshift". The light received at point (2) is redshifted by an amount $\Delta\lambda/\lambda$ approximately equal to (gh/c^2). The wavelength ratio $\Delta\lambda/\lambda$ is commonly called "redshift" by astronomers.

This result obtained from the accelerating elevator (a) also applies to elevator (b), which is stationary in a gravitational field. When a light pulse propagates from point (1) to point (2) in elevator (b), the gravitational field causes a redshift in the spectrum of the light. The gravitational redshift $\Delta\lambda/\lambda$ is approximately equal to gh/c^2.

The quantity gh is called the "gravitational potential", which is commonly denoted by the Greek letter ϕ (phi). We use the symbol ϕ' so that we can reserve ϕ for the normalized gravitational potential. Thus the redshift $\Delta\lambda/\lambda$ produced by a gravitational field is approximately equal to ϕ'/c^2, where ϕ' is the difference in gravitational potential between points (1) and (2). In relativity calculations, normalized units are generally used. With normalized relativistic units, the gravitational potential ϕ' is divided by c^2 to obtain the normalized gravitational potential ϕ. Therefore, the redshift $\Delta\lambda/\lambda$ caused by a gravitational field is approximately equal to the difference in normalized gravitational potential ϕ between the transmitter and receiver of the light. The normalized gravitational potential difference ϕ is equal to gh/c^2.

Suppose we placed a clock at point (1), and time the emitted light pulses with the ticking of the clock, so that one pulse is emitted every microsecond. We find that the spacing between the light pulses received at point (2) is greater than one microsecond. To understand this, imagine that you are running away from a train of light pulses that is emitted by a fixed source. As you run, the distance between you and the light emitter increases, and so the number of light pulses along the transmission path increases with time. Consequently, the rate at which you receive pulses must decrease.

This reasoning shows that the acceleration of the elevator (a) causes a clock at point (1) to tick more slowly when observed at point (2). Similarly the gravitational field in elevator (b) causes a clock on the floor of the elevator at point (1) to tick more slowly when observed from point (2) at the ceiling of the elevator. When observed from the ceiling at point (2), the clock at point (1) on the floor appears to be slowed approximately by the factor $[1 - (gh/c^2)]$. This clock-rate factor can also be expressed as $(1 - \phi)$, where ϕ is the difference in normalized gravitational potential between the floor and the ceiling of the elevator.

Now, let us assume that a light pulse is emitted at the ceiling of the elevator and is observed at the floor. In this case, point (1) appears to be moving toward point (2). A Doppler wavelength shift is observed, but the spectrum is now shifted toward the blue end of the spectrum (toward higher frequency or shorter wavelength). Hence when light travels from the ceiling to the floor of the elevator, the spectrum experiences a "blue-shift" approximately equal to the normalized gravitational potential difference ϕ between the ceiling and floor of the elevator. The quantity ϕ is equal to gh/c^2.

Similarly, when a clock that is located on the elevator ceiling is observed from the floor of the elevator, the clock appears to run fast. The clock rate observed at the floor is approximately equal to the clock rate seen at the ceiling multiplied by (1 + ϕ), where ϕ is the difference in normalized gravitational potential between the floor and the ceiling.

The gravitational field of the sun causes a redshift in the sun spectrum, when observed from earth. On earth we are located at the ceiling of the elevator (further from the gravitational mass of the sun) and the light source is located at the floor of the elevator (on the surface of the sun). If we were stationed on the surface of the sun (which is difficult to achieve!) and observed a light emitted from earth, the spectrum of the light would be blue-shifted. Similarly, a clock on earth would appear to run faster when observed from the surface of the sun.

The above reasoning is similar to the analysis that Einstein made to determine the wavelength change and the change of clock rate that are caused by a gravitational field. The calculations that Einstein performed were approximate.

Yilmaz repeated this analysis using exact calculations that applied the principles of special relativity. From this exact analysis, Yilmaz derived the equations that specify the static solution of his theory of gravity. This exact analysis is discussed at the end of Chapter 9.

The Meaning of a Tensor

To investigate the concepts of general relativity, we need to understand the mysterious concept of a tensor. As we will see, the basic principle of the tensor can be easily explained.

Specifying a Vector

We start by considering the vector. A vector is represented by an arrow. Vector arrows were shown in Fig. 6-3 of Chapter 6 to describe the velocity of light and the velocity of the space ship. The velocity of the space ship is shown as a vector arrow, in which the direction of the arrow indicates the direction of the ship velocity and the length of the arrow is proportional to the value of the velocity. Since the space ship velocity is 60 percent of the speed of light, the vector arrow representing the space ship velocity is 60 percent as long as the vector arrow representing the velocity of the light pulse radiated from the earth.

A vector arrow can represent any quantity that has direction as well as amplitude. A vector can be used to represent such things as velocity, force, acceleration, or motion of a vehicle during a specified time interval.

For example, a sailor can estimate the position of his ship using "dead reckoning". Vector arrows are drawn on a chart, representing the ship motion while traveling at a given heading. The direction of each vector arrow is determined from the direction of the ship measured from the compass, and the amount of ship motion at that heading establishes the length of each vector arrow. The amount of ship motion is equal to the speed of the ship relative to the water, multiplied by the time period.

The navigator adds the vector arrows by drawing them head-to-tail on the chart, and thereby determines the total motion of the ship. This example shows how vectors can be used and how they are added together.

A vector can be specified by giving its components in different directions. The ship navigator can describe the ship motion by specifying the easterly and the northerly components of a vector arrow. If the ship is heading south, its "northerly component" is negative. An airplane moves vertically as well as horizontally. Therefore the velocity of an aircraft must be specified by the airplane navigator in three dimensions, which give the airplane velocity in the easterly direction, in the northerly direction, and in the vertical direction.

Figure 7-2 shows a general means of specifying the components of a vector in three dimensions. There are three axes labeled x, y, z, which are perpendicular to one another. We call these axes "rectangular coordinates". They were first used by the mathematician Rene Descartes (1596-1650) and so are called "Cartesian coordinates" by scientists. One often considers the z-axis to be vertical and the x, y axes to be horizontal, with the x-axis pointing in the east direction and the y-axis pointing in the north direction. However, any orientation of the axes can be used, provided that their relative orientation is maintained.

The components of the vector V_1 in Fig 7-2 are obtained by drawing the three dashed lines from the tip of the vector perpendicularly to the x, y, z axes. The points where these three dashed lines intersect the axes are labeled x_1, y_1, and z_1, which are called the "coordinates of the V_1 vector". We can draw vector arrows along the x, y, z axes, from the origin to the x_1, y_1, and z_1 intersection points. These vectors drawn along the axes are called the "components of the vector V_1". We can add these three

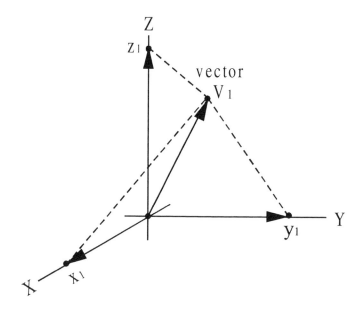

Figure 7-2: Rectangular coordinates of a vector

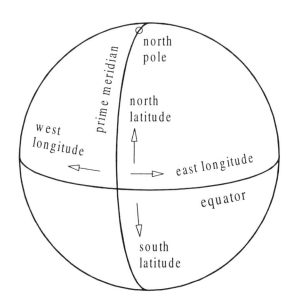

Figure 7-3: Latitude-longitude coordinates for specifying a location on the earth

component vectors together by displacing them without changing direction, and connecting them head-to-tail. The three component vectors will add together to form the total vector V_1.

We have seen how the components of a vector are specified in rectangular coordinates. One can also specify a vector in terms of spherical coordinates, which are related to the latitude-longitude coordinates for locating a point on the curved surface of the earth. Over a small region, the surface of the earth is approximately flat, and so we can find a location on a flat map. However, to specify location in an absolute sense relative to the earth, one must consider the curvature of the earth, and apply the angular measurements of latitude and longitude.

Latitude and longitude coordinates are defined in Fig. 7-3. For a point in the northern hemisphere, *north latitude* is measured in degrees from the equator in the northerly direction to the point of interest. For a point in the southern hemisphere, *south latitude* is measured in degrees from the equator in the southerly direction.

A *meridian* is a north-south line of constant longitude, which is a semicircle drawn from pole to pole. The *prime meridian* (or *Greenwich meridian*) is the line of constant longitude that passes through the old Royal Observatory in Greenwich, England. The prime meridian is defined as the line of zero longitude.

A point that is east of the prime meridian is specified by its *east-longitude* coordinate. East longitude is the angle measured in degrees along the equator in the easterly direction from the prime meridian to the meridian that passes through the point. A point that is west of the prime meridian is similarly specified by its *west-longitude* coordinate. By means of the latitude and longitude angles, one can locate a point on the surface of the earth in two dimensions.

This latitude-longitude approach can be generalized to obtain a three-dimensional specification by including the radial distance measured from the center of the earth. For example, the location of a satellite circling the earth can be specified in the following manner. Consider a vector drawn from the center of the earth to the satellite. The length of the vector is the radial coordinate r of the satellite. The latitude and longitude coordinates are measured at the point where the vector passes through the surface of the earth. The distance coordinate r, and the latitude, longitude angular coordinates, give a three-dimensional specification of the satellite location.

In relativity theory it is often convenient to specify spatial location in terms of spherical coordinates, rather than rectangular coordinates. With rectangular coordinates, three distances are specified to locate a point in three dimensions. These are distances measured along the x, y, and z axes. With spherical coordinates, the radial distance from the origin to the point of interest is specified, along with two angles. The angles are similar to the latitude and longitude coordinates for locating a point on the surface of the earth.

Tensor to Specify Stress within a Body

We can illustrate the concept of a tensor by considering the forces inside a mechanical body. A force exerted within a body is specified in terms of stress, which is the force applied per unit of area. Two directions are required to describe a stress. One direction gives the direction of the force, and the second direction gives the orientation of the surface to which the force is applied. The stresses exerted within a body are specified in terms of a mathematical concept called a "tensor".

As an example of the use of tensors, Fig. 7-4 shows the forces applied to a cube of material within a body. This body might be solid, liquid, or gas. These forces are expressed in term of the x, y, z axes. Each face of this small cube in Fig. 7-4 is considered to have unit area. *Force-per-unit-area* is called *stress*, and so the forces applied to the unit-area faces of the small cube are stresses.

These stresses are denoted in the form p_{jk}, where the subscript indices j, k can each represent x, y, or z. The first subscript index (j) describes the direction of the force. The second subscript index (k) describes the orientation of the face to which the force is applied. The orientation of a face is defined by a vector that is perpendicular to the face. We use the symbol "p" to represent a stress, because pressure is a typical example of a stress. Pressure, which is force per unit area, is specified with English units as pounds per square inch (psi).

Diagram (a) of Fig. 7-4 shows the three *compression stresses* that are applied at three faces of the cube, and are denoted p_{xx}, p_{yy}, and p_{zz}. Stress p_{xx} is exerted on the cube in the x-direction at a face that is perpendicular to the x-direction. Stress p_{yy} is exerted in the y-direction at a face that is perpendicular to the y-direction. Stress p_{zz} is exerted in the z-direction at a face that is perpendicular to the z-direction.

The three faces of the cube, which are specified by the second subscript index, are labeled the *x-face*, the *y-face*, and the *z-face*. The x-face is perpendicular to the x-direction, the y-face is perpendicular to the y-direction, and the z-face is perpendicular to the z-direction.

Diagram (b) of Fig 7-4 shows the six *shear stresses* that are applied to the faces of the cube. A shear stress is applied parallel to a face. There are two shear stresses applied to the x-face of the cube, which are denoted p_{yx} and p_{zx}. Stress p_{yx} is applied to the x-face in the y-direction, and stress p_{zx} is applied to the x-face in the z-direction. Similarly, the two shear stresses applied to the y-face are denoted p_{xy} and p_{zy}, and the two shear stresses applied to the z-face are denoted p_{xz} and p_{yz}.

Therefore nine separate stresses are required to describe the internal forces within a mechanical body. These stresses are arranged as follows in a 3-by-3 array, which is called a *matrix*

$$\begin{vmatrix} p_{xx} & p_{xy} & p_{xz} \\ p_{yx} & p_{yy} & p_{yz} \\ p_{zx} & p_{zy} & p_{zz} \end{vmatrix}$$

The elements are arranged in this array according to the following rule: the first subscript index indicates the row of an element, and the second index indicates the column. In the first row the first index is x; in the second row the first index is y; and in the third row the first index is z. Similarly the second index is x in the first column; the second index is y in the second column; and the second index is z in the third column.

The compression stresses, which were shown in diagram a of Fig 7-4, are denoted p_{xx}, p_{yy}, and p_{zz}. These compression stresses are located on the diagonal of the matrix, which extends from the upper left to the lower right. These three elements are called the "diagonal elements" of the matrix. The other elements (the shear stresses in this case) are called the "nondiagonal elements". In some matrices, the nondiagonal elements are all zero, and the matrix is called a "diagonal matrix".

This matrix is defined as the variable p_{jk}, which is called a *tensor*. The subscripts j, k are called indices. The variable p_{jk} represents these nine stresses in a general form. The indices j, k can each represent x, y, or z.

In summary, a tensor with two indices is required to define a stress variable, because stress has two independent directions. The first index specifies the direction of the force, and the second index specifies the orientation of the surface to which the force is applied. A surface is perpendicular to the direction specified by the second index.

It is often convenient to describe the x, y, z rectangular coordinates by numbers, where x is called x_1, y is called x_2, and z is called x_3. Hence the x index of a stress component is replaced by 1, the y index is replaced by 2, and the z index is replaced by 3.

7. General Relativity 99

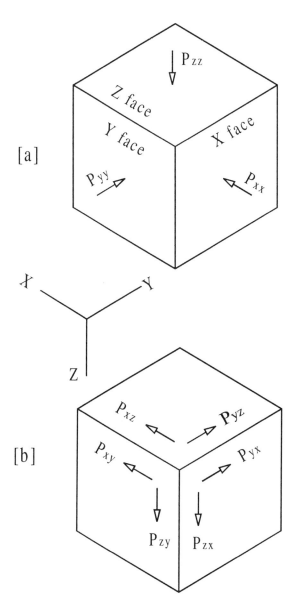

Figure 7-4: Internal stresses within a mechanical body: [a] compression stresses; [b] shear stresses.

In this form the stress tensor matrix is expressed as

$$\begin{vmatrix} p_{11} & p_{12} & p_{13} \\ p_{21} & p_{22} & p_{23} \\ p_{31} & p_{32} & p_{33} \end{vmatrix}$$

In the stress tensor p_{jk}, the subscript indices j, k now take the values, 1, 2, and 3. For elements in the first row, the first index is 1, in the second row the first index is 2, and in the third row the first index is 3. Similarly the elements are arranged in columns according to the second index.

The symbolism associated with tensors can be confusing. The reader should carefully study this description of the stress tensors so that it can provide a simple concept when a relativity tensor is discussed.

Tensors in Relativity Theory

In relativity theory, the space and time variables cannot be considered separately. Reality is described by *events*, where an event is specified by three spatial coordinates plus a time coordinate, and so has four dimensions. In relativity theory a vector has 4 components, and a tensor has 4x4 or 16 components.

In a relativity tensor, the x, y, z spatial coordinates are specified as x_1, x_2, x_3. Einstein denoted the time coordinate as x_4, but we use the alternative convention where the time coordinate is denoted x_0. Hence we represent a general relativity tensor by a 4 by 4 array of the form.

$$\begin{vmatrix} p_{00} & p_{01} & p_{02} & p_{03} \\ p_{10} & p_{11} & p_{12} & p_{13} \\ p_{20} & p_{21} & p_{22} & p_{23} \\ p_{30} & p_{31} & p_{32} & p_{33} \end{vmatrix}$$

The first row and the first column give the elements that are associated with the time coordinate.

This four-dimensional tensor is denoted $p_{\mu\nu}$. Greek-letter indices, such as μ, ν (called mu, nu), are used to describe a relativity tensor, where the indices take on the four values (0, 1, 2, 3). When three-dimensional spatial coordinates are specified, Roman-letter indices are used, such as j, k, which take on the three values (1, 2, 3). Hence, a relativity tensor is denoted in the form $p_{\mu\nu}$. Since the μ index can take on the four values (0, 1, 2, 3), and the ν index can take on the same four values (0, 1, 2, 3), the tensor $p_{\mu\nu}$ has 4x4 or 16 elements.

There are three different types of tensors, which are indicated by the indices as follows:

covariant tensor	$p_{\mu\nu}$	subscripted indices
contravariant tensor	$p^{\mu\nu}$	superscripted indices
mixed tensor	$p_{\mu}{}^{\nu}$	subscripted and superscripted indices

There are complicated theoretical issues associated with *covariant*, *contravariant*, and *mixed* tensors, but these need not concern us. We merely need to understand that there are three types of tensors, and that there are precise formulas that allow one to convert from one tensor form to another.

The variables representing distance and time in a tensor equation generally have superscripted (contravariant) indices, and so are represented as x^0, x^1, x^2, x^3. One must not confuse these superscripted indices with exponents, which would mean that the variable is being raised to a power.

When rectangular spatial coordinates are used, the symbols x^1, x^2, x^3 represent x, y, z, respectively. The symbol x^0 represents the time coordinate, which is equal to ct, and is the time t (in seconds) multiplied by the speed of light c. This is normalized time, which is denoted τ (Greek letter tau), and is the distance that light travels during the time interval t. Since the time variable x^0 (or τ) is equivalent to distance, it can be expressed in the same distance units as the spatial coordinates.

The Metric Tensor

The theoretical starting point of tensor analysis in general relativity theory is the *metric tensor*, denoted $g_{\mu\nu}$, which describes the *dimensional characteristics of space*. Einstein called this the *fundamental tensor*. For the examples that Einstein was able to consider, the metric tensors are all *diagonal*, which means that all of the elements of $g_{\mu\nu}$ that are not on the diagonal are zero. The matrix for a diagonal metric tensor $g_{\mu\nu}$ has the form:

$$\begin{vmatrix} g_{00} & 0 & 0 & 0 \\ 0 & g_{11} & 0 & 0 \\ 0 & 0 & g_{22} & 0 \\ 0 & 0 & 0 & g_{33} \end{vmatrix}$$

When the metric tensor is not diagonal, the equations derived from general relativity can have millions of terms, and so can only be solved with a computer. Since powerful computers were not available during Einstein's lifetime, Einstein had to restrict his studies to simple physical examples that resulted in *diagonal metric tensors*.

A region of space with no gravitational field is said to be *flat*. In a *flat* region of space, the metric tensor elements in rectangular coordinates are: $g_{00} = 1$; $g_{11} = g_{22} = g_{33} = -1$. Where the space is *flat*, the equations of special relativity apply exactly. Within our solar system, the gravitational field is weak, and so the metric tensor values are very close to these ideal flat-space values.

With this simple introduction to tensors, the reader should be able to understand our discussions of tensors without being confused by the strange tensor symbolism.

Relativistic Effects Produced by Gravity

Formulas for Relativistic Effects

Special relativity shows that velocity causes a clock to run slow and a measuring rod to contract. In general relativity, a gravitational field also causes a clock to run slow and a measuring rod to contract. In special relativity the speed of light is always constant, but in general relativity a gravitational field causes the speed of light to decrease. Because of the reduction of clock rate, excited atoms oscillate at lower frequency in a gravitational field, and this effect produces a spectral redshift. Let us examine these relativistic effects, which are produced by the gravitational field of a star.

Relativistic effects are determined from the elements of the metric tensor. We assume that the metric tensor $g_{\mu\nu}$ is diagonal, so that it has only four non-zero values. Ideally the three spatial elements g_{11}, g_{22}, g_{33} should be equal, when the tensor is expressed in rectangular coordinates. This condition does not actually hold for the Schwartzschild solution, but it does hold for the related *Isotropic* solution of the Einstein theory. When these three spatial elements are not equal, strange effects occur, because the spatial contraction due to gravity is different in the tangential and radial directions. Consequently the circumference of a circle is no longer equal to π times the diameter. These strange effects are discussed in Appendix G of *Universe* [1].

In this book we ignore this confusing issue, and assume that g_{11}, g_{22},

7. General Relativity 103

g_{33} are equal in rectangular coordinates. With this constraint, there are only two independent elements of the metric tensor that we must consider, g_{00} and g_{11}. The element g_{00} describes the temporal effects and g_{11} describes the spatial effects. Table 7-1 gives the formulas for the general relativity effects produced by a gravitational field. These formulas are derived in Chapter 8 of Universe [1].

Table 7-1: Relativistic Effect Caused by a Gravitational Field

Clock rate	$\sqrt{[g_{00}]}$
Spatial contraction	$1/\sqrt{[-g_{11}]}$
Speed of light	$\sqrt{[-g_{00}/g_{11}]}$
Wavelength ratio (λ'/λ)	$1/\sqrt{[g_{00}]}$

Table 7-2: Metric Tensor Elements for Einstein Schwartzschild Solution and Yilmaz Solution of a Star

	g_{00}	g_{11}
Schwartzschild	$1 - 2(m/r)$	$-1/g_{00}$
Yilmaz	$e^{-2m/r}$	$-1/g_{00}$

Formulas for the metric tensor elements g_{00}, g_{11} are given in Table 7-2 for the Einstein Schwartzschild solution and for the Yilmaz theory solution for a star. These formulas are expressed in terms of the normalized relativistic mass m of the star and the radial distance r from the center of the star.

The formula of g_{00} for the Yilmaz theory has an exponential form, which is probably strange to many readers. The reader can ignore this problem, because we will soon see the effect of this function in terms of graphical plots. Values of this exponential expression $e^{-2m/r}$ can be readily obtained from a scientific pocket calculator for particular values of the m/r ratio.

The normalized relativistic mass m is equal to MG/c^2, where M is the mass of the star expressed in conventional units, and G is the gravitational constant in Newton's law of gravitational force. Since m is proportional to the real mass M, it is convenient to regard m to be a "normalized mass". Defining a normalized mass in this manner is a convenient mathematical trick that simplifies the computations.

The mass M of our sun is 2.00×10^{30} kilogram, which means 2 followed by 30 zeros. Multiplying this by G/c^2 yields the following normalized mass m of our sun:

Normalized sun mass (m): 1.475 kilometers

Notice that normalized mass m has the units of distance. Since m and r both have distance units, the ratio m/r is a simple number without units.

Since the radius of our sun is 696,000 kilometers, the m/r ratio at the surface of the sun is 1.475/696,000, which is 1/472,000, or 2.12×10^{-6}. The expression 10^{-6} means 1/1,000,000, and so the m/r ratio at the surface of the sun is 2.12 parts per million. This is the maximum value of the m/r ratio in our solar system.

Within our solar system, the Schwartzschild expression for g_{00}, which is [1 - 2(m/r)], is close to unity, because m/r is much less than unity. Within our solar system the g_{00} value for the Yilmaz theory, which is $e^{-2m/r}$, is closely approximated by this value [1 - 2(m/r)] for the Schwartzschild solution.

Plots of Gravitational Effects

The values for g_{00}, g_{11} in Table 7-2 are applied to the formulas of Table 7-1 to obtain the specific formulas for the two solutions. These formulas are plotted in Figs 7-5 to 7-7 for both theories. The plots are shown with solid curves for the Einstein Schwartzschild solution, and with dashed curves for the Yilmaz single-star solution. Since the maximum value for m/r in our solar system is about 2 parts per million, only the first tiny bit of each plot applies to our solar system.

Figure 7-5 shows the speed of light for the two theories. For the Einstein theory, the speed of light goes to zero at m/r equal to ½. The point where m/r is ½ is the Schwartzschild limit. Since the speed of light is zero where m/r equals ½, it has been concluded from the Einstein theory that light cannot escape from a star if m/r at the stellar surface equals or exceeds ½. Such a star is called a *black hole*, and the spherical surface where r is equal to 2m is called an *event horizon*. Over this event-horizon surface, the speed of light should theoretically be zero. The radius of the event-horizon is called the *Schwartzschild radius*.

The dashed plot for the Yilmaz theory in Figure 7-5 does not go to zero at any point. Consequently the Yilmaz theory does not allow a black hole or an event horizon. The Yilmaz theory tells us that the black hole and event horizon concepts do not represent physical reality. This supports the position that was strongly asserted by Einstein.

7. *General Relativity*   105

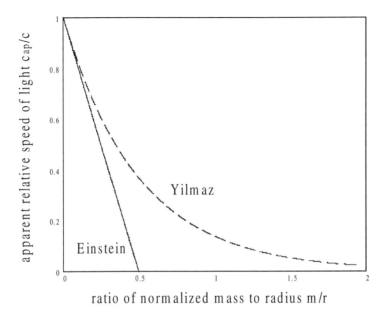

Figure 7-5: Apparent relative speed of light for Schwartzschild Einstein solution and for Yilmaz single-star solution.

The Schwartzschild analysis consists of two solutions: (1) the *interior* solution, which applies inside the star, and (2) the *exterior* solution, which applies in the vacuum of space outside the star. These two solutions are equated at the surface of the star. If the m/r ratio at the surface of the star exceeds ½, the interior solution yields an "imaginary" value for the pressure close to the center of the star. This physically impossible result shows that the Schwartzschild analysis does not apply if m/r exceeds ½.

The Schwartzschild analysis assumes a "static" model of the star, in which the dimensions of the star are constant. A real value for pressure inside the star can be achieved, with m/r greater than ½, if one assumes a "dynamic" model in which the star continually decreases in size. Under this condition, the star is theoretically surrounded by an event horizon of radius r equal to 2m. Inside this event horizon, the star should contract indefinitely until it shrinks to zero, thereby forming a singularity. Since the mass of the star does not change, the density of matter should become infinite as the size of the star shrinks to zero.

The Einstein theory predicts that when the m/r ratio at the surface of a star exceeds ½, the star starts to collapse, and it continues to collapse until the star shrinks to form a singularity.

However, closer examination of the Schwartzschild interior solution shows that another serious problem should occur when the m/r ratio at the surface of a star is greater than 4/9 but less than ½. Within this range of m/r, the pressure inside the star should become infinite over a spherical surface. This surface should progress from the center of the star to the circumference as m/r at the surface of the star increases from 4/9 to ½. Infinite pressure is not physically possible, and so it has been concluded that a star should become unstable and collapse into a black hole singularity when m/r at the star surface exceeds 4/9, rather than ½.

The difference between these m/r limits of 4/9 and ½ is important, because, as we will see, it greatly reduces the gravitational redshift that a star can produce according to the Einstein theory.

Figure 7-6 shows the contraction of dimensions and the reduction of clock rate that are predicted by the two theories. These are plots of the formulas in rows 2 and 3 of Table 7-1. The plots for both characteristics have the same shape, because g_{11} is equal to $-1/g_{00}$ for both theories.

Since the Einstein theory cannot allow a value for m/r in excess of 4/9, the formulas show that (for the Einstein theory) the minimum possible value for the spatial contraction ratio or relative clock rate is 1/3. This minimum value is indicated by a dot in the figure. In contrast, the spatial contraction ratio and relative clock rate for the Yilmaz theory (shown by the dashed plot) does not have a lower limit.

The effect of gravity on wavelength is shown in Fig. 7-7. The symbol λ represents the normal wavelength and λ' represents the observed wavelength. Figure 7-7 gives the plots of the wavelength ratio λ'/λ for the two theories calculated from the formula in the last row of Table 7-1. Since the maximum allowable value of m/r for the Einstein theory is 4/9, it can be shown from Tables 7-1 and 7-2 that the maximum value of the λ'/λ wavelength ratio for the Einstein theory is 3. This maximum value is shown as a dot in Fig 7-7. There is no fundamental upper limit to the λ'/λ wavelength ratio for the Yilmaz theory.

The observed wavelength λ' is defined as $(\lambda + \Delta\lambda)$, where $\Delta\lambda$ is the increase in wavelength above the normal wavelength λ. Redshift is defined as the increase in wavelength divided by the normal wavelength, and so represents the ratio $\Delta\lambda/\lambda$, which is equal to $(\lambda'/\lambda - 1)$. It can be seen from Fig. 7-7 that the maximum λ'/λ wavelength ratio due to gravity that can be predicted by the Einstein theory is 3. Hence the maximum $\Delta\lambda/\lambda$ gravitational redshift that can be predicted by the Einstein theory is 2. In contrast there is no fundamental upper limit to the gravitational redshift that can be predicted by the Yilmaz theory.

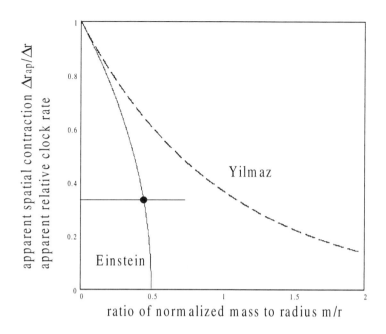

Figure 7-6: Apparent spatial contraction and relative clock rate for Einstein Schwartzschild solution and Yilmaz theory

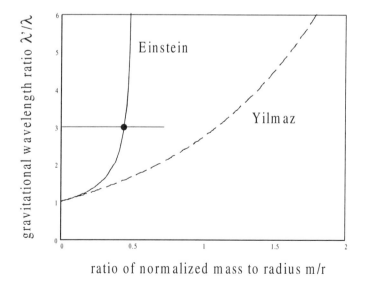

Figure 7-7: Wavelength ratio due to gravity for Einstein Schwartzschild solution and for Yilmaz theory

Quasars have been found with spectral redshifts as large as 5. Since the Einstein theory cannot predict a gravitational redshift greater than 2, it has generally been concluded that the quasar redshift must be produced by velocity. It is claimed that quasars must be receding at enormous velocities. They must be billions of light years away, and must radiate unbelievable amounts of energy.

In contrast, the Yilmaz theory can predict a gravitational redshift of 5 or greater. This indicates that the redshifts of all quasars can be caused primarily by gravity, and so quasars can be receding at much lower velocities. Thus the Yilmaz theory predicts that quasars are much closer than is normally assumed, and are radiating much less energy.

After quasars were discovered in 1963, the possibility that the quasar redshift might be caused by gravity was considered by astronomers. Chrandeskhar [23] applied the Einstein theory to analyze the mechanical stability of a star with an m/r ratio less than 4/9. He concluded that radial oscillations should occur inside the star unless the m/r ratio is much smaller than 4/9, and so the maximum value of gravitational redshift that a stable star can generate should be much less than 2. This analysis was a major factor leading to the general belief that gravitational redshift could not be an important factor in the quasar redshift.

This instability problem in a star calculated from the Einstein theory does not apply to the Yilmaz theory. The plot of spatial contraction in Fig. 7-6 for the Einstein theory has a very steep slope, even for values of m/r less than 4/9. This characteristic suggests that the Einstein theory should have mechanical stability problems for m/r values approaching 4/9. In contrast, the slope of the plot for the Yilmaz theory is gradual.

Therefore we conclude that the Yilmaz theory can readily explain the redshifts of all quasars as gravitational effects. The generally accepted belief that gravitational redshift cannot explain the extreme quasar redshift is based on analyses that applied the Einstein theory, which does not yield reliable results in an intense gravitational field.

The quasar enigma is discussed further in Chapter 8. The discussion will show that a second objection was raised against gravitational redshift, based on the lines in the quasar spectra. This second objection can also be dismissed.

Chapter 8

The Black Hole and the Quasar

The explanations of the *black hole* and the *quasar* given by Big Bang cosmologists are so bizarre they sound like science fiction. This chapter shows that they strongly conflict with observational evidence.

The Schwartzschild solution to the Einstein general theory of relativity characterizes the gravitational field of a star. The *black hole* concept evolved from an anomaly in the Schwartzschild solution, which is called the Schwartzschild singularity. Einstein always considered this anomaly to be non-physical, and we concur with Einstein. The discussion shows that the black hole concept is the consequence of a mathematical flaw in the Einstein theory. Although the Einstein theory gives accurate answers in the weak gravitational fields of our solar system, it does not give reliable results under the intense gravitational fields associated with the Schwartzschild singularity.

The Yilmaz theory predicts that the large redshift of a quasar can be readily explained as the effect of an intense gravitational field. It is certainly not a Doppler effect caused by a receding velocity. This interpretation of the quasar spectrum indicates that quasars are much closer than is generally believed, and are radiating much less energy.

The chapter discusses the observational evidence concerning quasars that was gathered by Halton Arp. His evidence proves that the redshift of a quasar cannot be a measure of its velocity. Arp has also shown that some galaxies are physically associated with other galaxies of appreciably lower redshift. Hence these galaxies must have intrinsic redshift that is not related to galaxy velocity. It is difficult to explain the intrinsic redshift of a galaxy as a gravitational effect, because a large gravitational redshift requires a highly compact body. However, we will see that the Marmet redshift theory shows that a dense cloud of hydrogen gas could produce the observed intrinsic redshift of a galaxy.

The Physically Impossible Black Hole

The black hole concept was derived from the Schwartzschild solution of the Einstein theory, which describes the gravitational effects of a star. As explained in Chapter 7, the black hole is predicted by the Einstein theory, but is inconsistent with the Yilmaz theory. The Yilmaz theory does not allow either a black hole or an event horizon. We have also seen that Einstein strongly opposed the black hole, because this concept requires the star that forms a black hole to collapse indefinitely until it becomes a singularity having essentially infinite density of matter. Einstein opposed this singularity prediction because it drastically violates our laws of physics.

Astronomical studies have shown that some stars in our Milky Way galaxy, called *white dwarf stars* and *neutron stars*, have suffered gravitational collapse to produce an extremely high density of matter. Consequently the postulate of infinite mass density inside a black hole might appear to be acceptable. Nevertheless our laws of physics still apply within the extremely compact white dwarf and neutron stars, and real matter still exists. *In contrast, real matter cannot exist within a physically impossible black hole, and our established laws of physics do not apply.* The following discussion examines the extremely compact white dwarf and neutron stars.

As explained by Goldsmith [14] (pps. 176-199), the primary nuclear reaction in a star is the conversion of hydrogen into helium. The heat generated by this reaction provides internal pressure that offsets the force of gravity. Our sun has been converting hydrogen into helium for five billion years, and will continue to do so for another five billion years. At the end of that period, the sun will swell to form a red giant, with a surface reaching half way to the earth. When all of the hydrogen is converted into helium, gravity will cause the sun to collapse, creating a temperature at the core of 100 million degrees Celsius. At that point a new fusion process will occur in which helium is converted into carbon.

The fusion of helium into carbon will continue for another billion years. When the helium is completely converted into carbon, nuclear fusion will cease, and gravity will compress the sun to an enormous density of matter. Our sun will become a *white dwarf*. It will shrink to the size of the earth, glowing white hot from the energy released by gravity. Finally when the sun can collapse no further, it will cool to become a black dwarf, the dead ember of a once brilliant star.

Goldsmith [14] (pps. 176-199) describes this aging process predicted for our sun. He reports that 99.9 percent of all stars end their lives in this manner as white dwarfs, which gradually fade into darkness as black dwarfs. However, one star in 1000 ends its life in a much more dramatic manner as an exploding supernova.

Matter within a white dwarf star is extremely dense, about 2 tons per cubic centimeter, which is one million times denser than our sun is today. Further gravitational collapse is prohibited in a star of this size by the Pauli exclusion principle, which states that electrons can only occupy specific energy states. Force from the electrons keeps the star from collapsing further.

On the other hand, if the star has sufficient mass, nuclear reaction does not stop with the helium converted to carbon. Further gravitational compression takes place, which achieves the temperatures required to fuse carbon with helium to form oxygen, and oxygen with helium to form neon. Carbon is fused with carbon to form silicon, and silicon is fused with silicon to form nickel, cobalt, and iron. Then the fusion stops, because the formation of heavier elements does not release energy; it absorbs energy.

When the nuclear fuel is depleted, massive gravitational collapse suddenly occurs. The pressures become sufficient to force the electrons into the protons to form neutrons. The atomic nuclei collapse and the star consists entirely of neutrons in direct contact with one another. The star is called a neutron star. Further collapse is prohibited by the Pauli exclusion principle applied to neutrons, which states that two neutrons cannot occupy the same state.

When an electron combines with a proton to form a neutron, a neutrino is released. This is an extremely small particle traveling at essentially the speed of light. It normally passes freely through a body, and only rarely interacts with matter. Most of the neutrinos falling on the earth pass directly through it much more easily than light passes through our atmosphere.

Since the matter in the collapsing neutron star is enormously dense, many of the neutrinos are absorbed by the outer layers of the collapsing star. The energy from the neutrinos blows the star apart in a tremendous explosion, which is called a *supernova*. For a brief period the star shines with the brightness of many millions of stars. The neutrinos react with the matter to form elements heavier than iron. These new elements, along with the other elements that were created earlier within the star, are scattered by the explosion as dust particles, which are spread throughout the galaxy. It is these dust particles from ancient exploding supernovas that supplied the solid material from which our earth was formed.

The theory indicates that a neutron star should be left after a supernova explosion. Astronomers have concluded that pulsars, which display radio waves in a series of short pulses, are rapidly spinning neutron stars.

The density of a neutron star is enormous: 300 million tons per cubic centimeter. Suppose that we use earth-moving equipment to dig a hole that is 1500 ft deep, over an area of 50 acres, to cover a square area 1500 ft on a side. We put the material dug from the hole into a super compactor, which compresses all of the material into a volume of one cubic centimeter. This cube would have the density of a neutron star.

Although the density of a neutron star is unbelievably high, it is still finite. Matter still exists as matter, and our laws of physics still apply. The Pauli exclusion principle prohibits further gravitational collapse.

If a neutron star has about 8 times the mass of the sun, it would have a normalized mass m of about 12 km, and a radius of about 24 km. Consequently the m/r ratio at its surface would be about ½. According to the black-hole theory, this extremely dense neutron star must shrink indefinitely until it forms a singularity of zero diameter and infinite density of matter.

Some theoreticians have argued that the mass density of the star inside a black hole does not actually become infinite. Nevertheless, the black hole concept requires, as a minimum, that the density of matter inside a black hole must be many, many times greater than the fantastically high density of a neutron star. Our laws of physics still apply within a neutron star, and the Pauli exclusion principle still holds. However physical matter (as we know it) cannot exist inside the event horizon of a black hole. It seems inconceivable that Albert Einstein would ever have accepted the black hole as representing physical reality.

Cosmologists have taken Einstein's theory as proof that the center of a black hole achieves a mass density that is very much greater than the extremely high density of the neutron star. Under this condition the laws of physics break down. In a black hole, the Pauli exclusion principle is violated, and real matter ceases to exist. *Einstein never accepted this claim that his theory violates the Pauli exclusion principle.*

The *black hole solution* to the Einstein general theory of relativity is strong evidence that the Einstein gravitational field equation has a fundamental mathematical weakness. This is corrected by the Yilmaz theory, which has modified the Einstein gravitational field equation. With this modification, the Yilmaz solution does not yield a physically impossible black hole.

The Quasar Enigma

Cause of the Quasar Redshift

Quasars are star-like objects that have extremely large redshifts in their spectra. They appear as points of light to the telescope. The first quasars had strong emission at radio frequencies. Consequently they were named "quasi-stellar radio objects", and were given the acronym "quasar". Later studies showed that only about 10 percent of quasars are radio sources, and so these objects are now called "quasi-stellar objects", often abbreviated QSO. Nevertheless the original "quasar" name is still widely used.

It is generally believed that the very large redshift in the quasar is a Doppler effect caused by a very high receding velocity. With such a large velocity, a quasar must be extremely distant, billions of light years away. At such a great distance it must radiate an enormous amount of energy for us to see it. It is claimed that a quasar typically radiates ten or more times the total output from our Milky Way galaxy.

The light output from many quasars can vary with time by a factor of two or more, and the period of variation can be as short as a day. This indicates that the dimensions of many quasars are small, and some quasars can be no larger than our solar system. How can an object that is about the size of a star radiate 10 times the total energy emitted from our enormous Milky Way galaxy, which is one hundred thousand light years across?

This enigma can be explained by assuming that an intense gravitational field is causing the large quasar redshift. As explained in the website *Addendum* [2], Appendix G, this alternative was rejected by astronomers soon after quasars were discovered for the following two reasons:

(1) Analyses using the Einstein theory indicated that a star should become mechanically unstable before it has a sufficient mass-to-radius ratio to achieve a large gravitational redshift.

(2) Strong "forbidden" spectral lines of oxygen and neon are observed in the quasar spectra. These forbidden lines are not encountered on earth, and are only observed in radiation from gaseous nebulae, which are huge clouds of very thin gas illuminated by stars. Based on the theory of forbidden lines, which evolved from studies of gaseous nebulae, it was concluded that the atmosphere of a star that

exhibits large gravitational redshift would be far too small to generate forbidden spectral lines.

As was explained in Chapter 7, argument (1) is refuted by the Yilmaz theory. The instability predicted by the Einstein theory at large mass-to-radius values is merely the result of a flaw in the Einstein theory that is related to the black hole. Since the Yilmaz theory does not have this difficulty, it can predict sufficient gravitational redshift to explain the redshifts of all quasars.

Appendix G of the website *Addendum* [2] analyzes the issue of forbidden spectral lines. These are spectral lines of oxygen and other elements that have low probability of being emitted. They are not produced on earth, and, except for quasar spectra, have only been observed in the radiation from gaseous nebulae. A gaseous nebula is a huge cloud of very thin ionized gas that is heated to a temperature of about ten thousand degrees Celsius by the radiation from stars. The gas density is usually less than 100,000 electrons per cubic centimeter. For comparison, the electron density of a high vacuum produced on earth exceeds this density by a factor of 10 billion.

From observational studies of gaseous nebulae, combined with theoretical analyses, an extensive theory of forbidden spectral lines was developed. This theory was applied to quasar spectra soon after the extreme quasar redshift was discovered in Feb. 1963 by Maarten Schmidt and Jesse L. Greenstein. A star having high gravitational redshift must be compact, and so Greenstein and Schmidt concluded, in their classic paper on quasars [22], that the atmosphere of such a star would be far too small to generate the forbidden spectral lines that are observed in quasar spectra. Therefore they stated that the extreme redshift of the quasar could not be caused by gravity. It must be a Doppler effect produced by an extremely high receding velocity.

In this paper [22], Greenstein and Schmidt studied two quasars. The one with the greater redshift (0.367) was 3C48. The analysis of this quasar is evaluated in *Addendum* [2], Appendix G. Greenstein and Schmidt concluded that quasar 3C48 is 3.6 billion light years away. Based on the relative amplitudes of the forbidden spectral lines, they estimated that the thin gas that is generating these lines has an electron density of 30,000 electrons per cubic centimeter. To produce the measured spectral line intensities, a huge gas volume is required, the size of a sphere 50 light years in diameter.

However, this huge gas volume is inconsistent with observations that the brightness of 3C48 varies with time. The power received from 3C48 varied by a factor of 1.4 over a period of 600 days. With such a rapid brightness variation, the gas cloud cannot be 50 light years in diameter. The diameter must be smaller by at least a factor of 10, and so the gas volume must be at least a factor of 1000 times smaller. This finding indicates that the theory of forbidden spectral lines, which was derived from studies of gaseous nebulae, cannot explain the forbidden lines in the 3C48 quasar spectra, even in a crude sense.

In time, quasars were discovered with much larger redshifts and with much more rapid variations of quasar brightness. Quasar brightness has been found to vary by about a factor of 2, over periods of months, weeks, days, and even hours. It is obvious that the theory of forbidden spectral lines, which evolved from studies of huge gaseous nebulae, does not apply at all to these rapidly varying quasars.

We conclude that the theory of forbidden spectral lines derived from gaseous nebulae is grossly inconsistent with the forbidden lines that are observed in quasar spectra. Consequently a different explanation for these forbidden quasar lines is needed.

Let us reconsider the gravitational redshift possibility. If a quasar is a massive compact object, its atmosphere would experience an extreme gravitational field with a high gravitational gradient. The quasar would probably have a very strong magnetic field, which would generate large electric currents in its ionized atmosphere. The physical conditions of gasses in this atmosphere would be radically different from those encountered on earth or in a gaseous nebula. Hence forbidden spectral lines may well be emitted in this atmosphere at an electron density that is very much greater than is observed in gaseous nebulae. These issues are discussed further in Appendix G of the website *Addendum* [2].

Therefore we conclude that forbidden spectral lines in quasar spectra do not refute the possibility that the quasar redshift is caused by gravity. In fact, this possibility may well provide the only viable explanation of these forbidden spectral lines, when we consider the rapid brightness variations of many quasars.

We have seen that the Yilmaz theory can explain the redshifts of all quasars as gravitational effects, and does not exhibit the mechanical instability that is predicted by the Einstein theory. The presence of forbidden spectral lines in quasar spectra is consistent with the gravitational redshift explanation. Hence we have refuted both arguments that have been raised against gravitational redshift as an explanation for the quasar redshift.

Quasar Observations of Halton Arp

Starting in the 1960's, astronomer Halton Arp has amassed a large body of observational evidence demonstrating that quasars are relatively close objects, and that the redshift of a quasar does not characterize its velocity. This material is described in his most recent book, *Seeing Red* [5], which the author highly recommends.

Arp has presented strong statistical evidence of quasars that are directly associated with galaxies having much smaller redshift. When the image of one quasar is close to that of a galaxy with a much smaller redshift, it is not unreasonable to assume that this might be a chance relationship. The quasar might be billions of light years beyond the galaxy. However when two or more quasars appear to be close to a galaxy of much smaller redshift, a chance relationship is highly unlikely. We ask the question, "What is the probability that the images of two or more quasars could be this close to an arbitrary direction in space?" Statistical arguments provide strong evidence that the close angular relationship between the images of the galaxy and the quasars indicates a close physical relationship between them.

Arp photographed three quasars that appear to be in the outer fringe of galaxy NGC 3842. It is theoretically possible that this is a chance relationship; that the quasars actually lay far beyond the galaxy. However, the probability that three quasar images would fall this close to an arbitrary direction in space is about one in a million. As *Universe* [1] shows in Chapter 12, Arp has found several cases like this. *The possibility is essentially zero that all of these observations could be accidental, which show associations between quasars and galaxies having much smaller redshifts.* Other statistical examples of this sort are given in Arp's book, *Seeing Red* [5]. This book also describes other types of statistical evidence to support the conclusion that quasars are relatively close objects..

Besides this very strong statistical evidence, Arp has found many examples of filament structures that directly connect quasars to galaxies of much lower redshift. These are described in his book, *Seeing Red* [5]. Arp has presented many examples showing that quasars seem to have been ejected from galaxies having much smaller redshifts. The galaxies for such cases are usually Seyfert galaxies, which have very bright nuclei that emit high x-ray and infrared radiation. Luminous jets related to quasars are often observed projecting from a galaxy in opposite directions. This suggests that quasars are often ejected from galaxies, possibly because of supernova explosions.

The book, *Seeing Red*, is filled with similar evidence. The scientific case that Arp has amassed is enormous. Nevertheless, he continues to experience great opposition to the presentation of his results in astronomical conferences and journals. This reinforces the conclusion that there is a severe lack of scientific objectivity in astronomy today. Scientific evidence has been replaced by consensus, and so astronomical science is deteriorating into mythology.

Galaxies with Intrinsic Redshift

In his book *Seeing Red* [5], Arp has presented numerous examples of small galaxies that are physically tied with filament structures to larger galaxies having smaller redshifts. The evidence suggests that the small galaxies have been ejected from the larger galaxies. This indicates that a galaxy can have an "intrinsic redshift" that is not related to its velocity.

We have seen that the Yilmaz theory can explain the intrinsic redshift of a quasar as an effect produced by an intense gravitational field. To produce gravitational redshift, a quasar must be a very compact body. Since quasars look like points of light to the telescope, it is reasonable to assume that these objects are very compact.

On the other hand, gravitational redshift cannot readily explain the intrinsic redshift of a galaxy, because it has an extended image. The measured redshift of a galaxy is constant over its image. To assume that an excess redshift of a galaxy is caused by gravity would seem to require that the galaxy consist of many identical compact bodies with equal gravitational redshift.

The cosmological theory of Paul Marmet [18], which was discussed in Chapter 3, provides an explanation for the intrinsic redshift of galaxies. An analysis of the Marmet effect is given in Appendix G of the website *Addendum* [2]. Marmet has shown that a photon of light loses energy when it collides with a gas molecule, and this produces a redshift, but the direction of the light does not change. This redshift effect cannot be observed in the earth's atmosphere, because the density of gas is too high. It cannot be observed in the laboratory, because it requires too long a path to create a measurable effect.

However, Marmet has shown that his effect can explain the variation of redshift across the disk of the sun for radiation emitted by the sun. The redshift of radiation from the limb of the sun is greater than that from the center of the sun. Thus, Marmet has experimental evidence to support his theory.

Most of the gas in space is probably hydrogen. Marmet has shown

[19] that hydrogen gas consists of atomic hydrogen H and molecular hydrogen H_2. In its atomic form, hydrogen is easily ionized, but the molecular form H_2 is extremely stable. Ionized hydrogen can be observed with radio astronomy, but atomic hydrogen is very difficult to detect in space. There is probably much more molecular hydrogen in space than atomic hydrogen.

The Marmet redshift effect is discussed in Chapter 3, and is analyzed in the website *Addendum* [2], Appendix G. In *Seeing Red* [5], Arp gives evidence of intrinsic galaxy redshifts as large as 6000 km/sec. Based on the data given in Chapter 2, we can explain an intrinsic redshift equivalent to 6000 km/sec by a hydrogen gas cloud that is 10 thousand light years thick and has a density of 1000 hydrogen atoms per cubic centimeter. Therefore the Marmet redshift effect can explain the intrinsic galaxy redshifts observed by Halton Arp using reasonable physical assumptions.

Since the Marmet redshift effect can explain the intrinsic redshifts of galaxies, we should also consider it to be a possible explanation for the intrinsic redshifts of quasars. Thus we have two promising explanations for the extreme redshifts of quasars: gravitational redshift and the Marmet redshift effect. In any case, there is strong evidence that the Doppler velocity effect is only a small component of the quasar redshift.

Other Quasar Enigmas

Halton Arp has shown in *Seeing Red* [5] (p. 56, Fig. 2-18) that quasar 3C48 (with a redshift of 0.367) is surrounded by a fuzzy galaxy-like body. In a personal communication, Halton Arp reported that the redshift of this fuzzy body is almost the same as quasar 3C48. The redshift difference is only 0.001.

There does not seem to be a reasonable explanation for this redshift observation. The Marmet redshift effect does not seem to explain the forbidden lines in the 3C48 spectrum. Why should the quasar and the galaxy have almost the same redshift? This example indicates that much more study of the quasar is needed before we can obtain an explanation that agrees with all of the evidence.

Chapter 9

Application of Einstein and Yilmaz Theories

The Einstein Gravitational Field Equation

This Chapter shows how the Einstein and Yilmaz theories are applied. We start with the Einstein theory, outlining its computational steps in order to explain the physical meaning of the variables that are being calculated. We will not be concerned with the detailed mathematical calculations. These calculations employ standard calculus techniques, which can be readily implemented by one who is skilled in the use of calculus.

The Einstein theory is applied to a physical model by solving its gravitational field equation, which is:

$$G_\mu^{\,\nu} = -8\pi T_\mu^{\,\nu} \quad \text{(Einstein gravitational field equation)}$$

The tensor $T_\mu^{\,\nu}$ is the energy momentum tensor, which describes the characteristics of matter. The Einstein tensor $G_\mu^{\,\nu}$ describes the curvature of space, and takes the place of gravitational force in the gravitational theory of Isaac Newton.

This tensor formula actually represents 16 separate equations. The individual equations are obtained by setting μ equal to (0, 1, 2, 3) and by setting ν equal to (0, 1, 2, 3). For example, when μ and ν are both 0, the formula shows that $G_0^{\,0}$ is equal to $-8\pi T_0^{\,0}$; when μ is 1 and ν is 3, it shows that $G_1^{\,3}$ is equal to $-8\pi T_1^{\,3}$; etc. In this manner, 16 separate equations can be derived from this tensor formula. Because of tensor symmetry, 6 of these 16 equations are redundant. For example, $G_1^{\,3}$ can be calculated directly from $G_3^{\,1}$, etc. With this symmetry, only 10 of the 16 equations are independent.

Therefore the tensor gravitational field equation of the Einstein theory *represents ten independent equations.* There are precise calculus formulas that allow one to compute the Einstein curvature tensor $G_\mu^{\ v}$ from the metric tensor $g_{\mu v}$. The energy-momentum tensor $T_\mu^{\ v}$ is derived from the characteristics of the physical model.

To obtain his energy momentum tensor, Schwartzschild considered a simple physical model for a star. He assumed that the star has no viscous (or shear) forces, and has a constant density of matter. Since a star is gaseous, the assumption of constant mass density is inaccurate, but was needed to achieve equations that could be solved analytically.

The energy momentum tensor that Schwartzschild derived for a star is diagonal, which means that the 12 nondiagonal elements are all zero. There are four nonzero elements, which are $T_0^{\ 0}$, $T_1^{\ 1}$, $T_2^{\ 2}$, and $T_3^{\ 3}$. The three elements $T_1^{\ 1}$, $T_2^{\ 2}$, $T_3^{\ 3}$ are each equal to -p, where p is the pressure within the star. The element $T_0^{\ 0}$ is equal to σ (Greek letter sigma), where σ is the density of matter within the star. Schwartzschild assumed that the density σ is constant, and that the pressure p varies with radial distance from the center of the star.

These pressure and density parameters are expressed in normalized relativistic units. When conventional units are used for pressure and density, the parameters are multiplied by G/c^2, just as is done in calculating the normalized mass m.

These elements of the energy-momentum tensor were applied to the gravitational field equation (shown above) to obtain the following elements for the Einstein tensor: $G_0^{\ 0}$ is $-8\pi\sigma$, and $G_1^{\ 1}$, $G_2^{\ 2}$, $G_3^{\ 3}$ are each equal to $8\pi p$. All other elements of the Einstein tensor are zero.

The Schwartzschild analysis consists of two solutions: the interior solution, which applies inside the star, and the exterior solution, which applies outside the star. The above values for the Einstein tensor are for the interior Schwartzschild solution. In the vacuum of space outside the star, the energy-momentum tensor $T_\mu^{\ v}$ is identically zero. Consequently, in the exterior Schwartzschild solution, all elements of the Einstein tensor $G_\mu^{\ v}$ are zero.

There are precise but very complicated calculus equations that allow one to calculate the elements of the Einstein curvature tensor $G_\mu^{\ v}$ from those of the metric tensor $g_{\mu v}$. Schwartzschild had to solve these equations backward in order to determine the $g_{\mu v}$ metric tensor elements that would produce the desired elements of the Einstein tensor $G_\mu^{\ v}$.

Schwartzschild achieved this backward calculation by assuming general formulas for the elements of the metric tensor $g_{\mu v}$. The metric

9. Application of Einstein and Yilmaz Theories

tensor was specified to be diagonal, and so had four unknown metric tensor elements. From these general formulas, he calculated the corresponding expressions for the elements of the Einstein tensor $G_\mu{}^v$. He set these equal to the desired values. From the resultant equations he was able to calculate the actual formulas for the elements of the metric tensor.

This Schwartzschild analysis assumed a simple physical model of a single star. An essential requirement of this analysis is that the metric tensor must be diagonal. If it is not, the equations for calculating the elements of the Einstein tensor $G_\mu{}^v$ from the metric tensor $g_{\mu v}$ can have millions of terms, and so cannot be solved analytically.

It was not until the 1960's that computers became available to solve the Einstein gravitational field equation for a complex physical model. However, even with a powerful computer, the solution of the Einstein gravitational field equation for a general model is very difficult, because the equations can yield millions of terms, and they must be solved backward. A complicated iterative computer program is needed to achieve this backward calculation.

The iterative program starts with approximate values for the metric tensor elements, and from these it computes the corresponding elements of the Einstein tensor. The computed elements of the Einstein tensor are compared with the desired elements, and the differences are used to change the metric tensor elements in such a way as to reduce the differences. The program cycles through this process until the computed elements of the Einstein tensor match the desired elements. The iterative computer program may perform billions of computations before a solution is obtained. Sophisticated mathematical techniques are generally required to achieve an iterative computer program that converges to a solution.

Since the 1960's, hundreds of mathematicians, physicists, and engineers in academic positions have devoted their careers to the task of solving the Einstein gravitational field equation on the computer. Highly advanced analytical procedures have been developed to achieve these computer solutions.

Albert Einstein displayed great genius in developing his theory of relativity. However, a similar genius is not needed to apply the Einstein theory. All of the mathematical relations associated with the Einstein gravitational field equation are precisely specified. It takes great skill in mathematical computer techniques to solve these equations on a computer. Nevertheless, these computer skills do not place a theoretician who is solving the Einstein equations in the same intellectual category as

Albert Einstein.

The Yilmaz Gravitational Theory

To simplify his theory, Yilmaz uses as his energy-momentum tensor the variable $\tau_\mu^{\ \nu}$, which is equal to $4\pi T_\mu^{\ \nu}$. (Do not confuse this tensor with the normalized time parameter τ, which represents ct.) In terms of this tensor, the Einstein gravitational field equation becomes:

$$G_\mu^{\ \nu} = -2\tau_\mu^{\ \nu} \qquad \textit{(Einstein gravitational field equation)}$$

The gravitational field equation for the Yilmaz theory has an additional tensor $t_\mu^{\ \nu}$, which is called the *stress-energy tensor for the gravitational field*. The gravitational field equation for the Yilmaz theory is

$$G_\mu^{\ \nu} = -2(\tau_\mu^{\ \nu} + t_\mu^{\ \nu}) \qquad \textit{(Yilmaz gravitational field equation)}$$

Yilmaz calls his energy-momentum tensor ($\tau_\mu^{\ \nu}$) the *stress-energy tensor for matter*.

As we have seen, the Einstein theory is applied by solving its very complicated gravitational field equation. The Yilmaz theory is very much easier to apply, because Yilmaz has derived general solutions to his gravitational field equation. Consequently the Yilmaz gravitational field equation does not need to be solved.

For nearly all practical applications, the static solution of the Yilmaz theory can be used. The static solution gives an exact answer only when the gravitational field does not vary with time. However, it gives an extremely accurate approximation provided that the velocities of the associated bodies are much less than the speed of light, a condition that is almost always satisfied.

The Yilmaz static solution is implemented by calculating the gravitational potential, denoted ϕ. For a single star having spherical symmetry (or any spherically symmetric body), the gravitational potential ϕ outside the body is simply

$$\phi = m/r$$

where m is the normalized relativistic mass of the star, and r is the radial distance measured from the center of the star. The density of the star can vary with radial distance from the center of the star, but must not vary with angle. If there are multiple bodies having spherical symmetry, the gravitational potential at a point outside all of the bodies is simply

$$\phi = (m_1/r_1) + (m_2/r_2) + (m_3/r_3) + \text{etc.}$$

Parameter m_1 is the normalized mass of body (1) and r_1 is the distance from the center of body (1) to the point of interest. The other quantities are defined accordingly. When the point of interest lies within one of the bodies, the term for that body is different. This issue is explained in Appendix B of *Universe* [1].

In the static Yilmaz solution, the metric tensor is always diagonal, which means that all elements of the metric tensor not on the diagonal are zero. In rectangular coordinates, the four diagonal elements are:

$$g_{00} = e^{-2\phi} \; ; \; g_{11} = g_{22} = g_{33} = -1/g_{00}$$

All other elements of the metric tensor are zero. For a single star, the gravitational potential ϕ is equal to m/r, and the formula for g_{00} is $e^{-2m/r}$. This formula was shown in Table 7-2 of Chapter 7, and was used to obtain the plots for the Yilmaz theory given in Figs 7-5 to 7-7.

These solutions for the Yilmaz theory are remarkably simple yet have broad applicability. The Schwartzschild solution applies only to a single body having a constant density of matter. However, the Yilmaz solution allows the density of matter to vary with radial distance from the center of the star. Besides (as was shown above) this single-star solution can be extended to characterize the gravitational effects of multiple bodies. It can be used to calculate the gravitational effects of all of the bodies of our solar system.

The general time-varying solution of the Yilmaz theory is analyzed in Chapter 5 and Appendix F of the website *Addendum* [2]. The time-varying Yilmaz solution is much more complicated than the static Yilmaz solution, but is still very much easier to apply than the Einstein theory. In the general time-varying solution of the Yilmaz theory, the gravitational potential ϕ is generalized to become the gravitational potential tensor ϕ_μ^ν. The gravitational potential ϕ is the "trace" of the tensor ϕ_μ^ν, which is the sum of the four diagonal elements of the tensor.

When the gravitational field varies slowly relative to the speed of light, the element ϕ_0^0 of the gravitational potential tensor ϕ_μ^ν is very much larger that all of the other elements, and so the other elements have negligible effect. The tensor ϕ_μ^ν reduces to the single element ϕ_0^0, which is equal to the trace ϕ of the tensor. Consequently, when the gravitational field varies slowly with respect to the speed of light, the general solution reduces to the simple static solution, in which the gravitational potential tensor $\phi_\mu{}^v$ is accurately represented by the gravitational potential ϕ.

Since the Yilmaz theory is very much easier to use than the Einstein theory, one might think that the Yilmaz theory would be strongly endorsed. This might be true if the Einstein general relativity theory were being applied to practical applications, but that is rarely the case. In artificial Big Bang studies, the great simplicity of the Yilmaz theory is a severe disadvantage.

If the Yilmaz theory is correct, the Einstein gravitational field equation must be wrong, and the sophisticated computer techniques that have been developed by hundreds of theoreticians since the 1960's to solve the Einstein theory are obsolete. They are not needed to apply the simple Yilmaz theory. Besides, the countless theoretical studies of the Big Bang theory are irrelevant.

The Gravitational Field Equations for the Einstein and Yilmaz Theories

Derivation of field equations. Chapter 7 showed how Einstein applied the equivalence principle to calculate the reduction of clock rate and the gravitational redshift that are produced by a gravitational field. Einstein used an approximation in this calculation. When Yilmaz examined this analysis, he realized that it could be implemented exactly. He applied an exact calculation, which used the Einstein formulas of special relativity. From this exact calculation, Yilmaz derived the metric tensor elements for the static solution of his theory. This derivation is discussed at the end of this chapter.

From the metric tensor elements of his theory, Yilmaz derived the corresponding gravitational field equation. Thus, Yilmaz derived his gravitational field equation in a rigorous manner. In contrast, Einstein used an intuitive approach to derive his equation. Yilmaz had the advantage of many years of research that had been performed in relativity theory, which gave him an understanding that was not apparent to Einstein during his pioneering investigation.

John A. Peacock [15] has written a lengthy book, called *Cosmological Physics*, which describes theoretical Big Bang research, including a discussion of relativity theory. He states (p. 19) that the Einstein gravitational field equation "cannot be derived in any rigorous

9. Application of Einstein and Yilmaz Theories   125

sense; all that can be done is to follow Einstein and start by thinking about the simplest form such an equation might take."

On pp. 26-27, Peacock [15] discusses possible "alternative theories of gravity". He asks, "How certain can we be that Einstein's theory of gravitation is correct?" He notes that the Einstein theory does not apply in sub-atomic scales, and goes on to say, "Apart from this restriction, there are no obvious areas of incompleteness. . . Nevertheless, . . it is possible that more accurate experiments will yield discrepancies. Over the years this possibility has motivated many suggestions of alternatives to general relativity."

Peacock is a strong supporter of the Big Bang theory, which is solidly tied to the Einstein theory. Nevertheless, even Peacock admits that the Einstein gravitational field equation was derived in an intuitive manner, and that a number of alternatives to that equation have been seriously considered by responsible scientists.

Einstein pseudo tensor. A crucial requirement that Einstein faced in developing his gravitational field equation is that all elements of his equation must be true tensors. As was shown in Chapter 6 in our discussion of Special Relativity, the "*Principle of Covariance*" must be satisfied in a relativity theory, so that the same laws of physics apply in all coordinate systems. To satisfy this principle when he generalized his relativity theory, Einstein constrained his gravitational field equation to consist only of tensors. Tensor theory provides a precise formula for transforming tensors into different coordinates. Hence if physical laws are expressed in terms of true tensors, the laws hold in all coordinate systems.

The Einstein gravitational field equation contains the Einstein tensor $G_\mu^{\ \nu}$, which describes the curvature of space, and the energy-momentum tensor, $T_\mu^{\ \nu}$, which describes the energy and stress for matter. Unlike the Yilmaz theory, the Einstein theory does not have a tensor to specify the energy and stress of the gravitational field.

Einstein searched for a tensor to characterize the energy of the gravitational field, but could not isolate a true tensor for this purpose. Einstein found what is called a "pseudo-tensor" to describe the energy of the gravitational field, which he denoted $t_\mu^{\ \nu}$. To distinguish this Einstein pseudo tensor from the corresponding true tensor of the Yilmaz theory, we denote the Einstein pseudo-tensor in bold letters with a strike-through mark as $\bcancel{t}_\mu^{\ \nu}$. Einstein believed that his pseudo tensor represented the "energy components of the gravitational field". However, it is not a true tensor, and so could not be used in the Einstein gravitational field equation.

Yilmaz has shown that $4\pi t_\mu^\nu$ is equal to $(-t_\mu^\nu + z_\mu^\nu)$, where $\mathbf{t}_\mu^\nu$ is the Einstein pseudo tensor, t_μ^ν is the Yilmaz stress-energy tensor for the gravitational field, and z_μ^ν is a non-tensor, which varies with the coordinate system. The non-tensor component z_μ^ν corrupts the Einstein pseudo-tensor $\mathbf{t}_\mu^\nu$ and keeps it from acting like a true tensor. Yilmaz was able to isolate the true tensor t_μ^ν that characterizes the energy of the gravitational field, which Einstein had sought.

The fact that the Einstein gravitational field equation lacks a tensor to specify the stress and energy of the gravitational field is a severe limitation. Chapter 10 discusses the implications of this deficiency.

The Einstein and Yilmaz theories both have energy-momentum tensors, which can be expressed in equivalent form as the tensor τ_μ^ν. However, the expressions for τ_μ^ν are not the same for the two theories, even when the physical model being analyzed is the same. This issue is explained in Chapter 10, and is related to the conservation of matter-plus-energy. This property is achieved by placing appropriate constraints on the energy-momentum tensor. It will be shown that the Yilmaz theory always achieves conservation of matter-plus-energy. However, the Einstein theory generally does not, and so the τ_μ^ν energy-momentum tensors for the two theories usually cannot be the same.

Uniqueness of Yilmaz Solution

A perplexing problem encountered in applying the Einstein theory is that it can yield many solutions for the same physical model. A general gravitational solution needs only six equations, because there are six degrees of freedom, corresponding to three linear motions and three rotational motions. Since the gravitational field equation represents ten independent equations, four of these equations are redundant. This redundancy results in contradictory solutions, which are a serious problem with the Einstein theory.

The Yilmaz theory does not have this problem, because Yilmaz has added theoretical constraints to his gravitational field equation to achieve a general solution that it unique. The constraints are called "gauge conditions" and "harmonic coordinates", which are highly desirable theoretically. With these constraints, the Yilmaz theory yields a general solution that is unique. The fact that these constraints result in a general solution proves that the constraints are valid. The Yilmaz theory is the only gravitational theory that has a general solution, and this property demonstrates that the Yilmaz theory has profound mathematical integrity.

Since the Yilmaz theory can yield only one answer for a given physical model, predictions derived from the Yilmaz theory are unique. Consequently these predictions should be taken seriously.

The Einstein theory can yield different solutions for the same physical model, depending on the mathematical constraints that are applied in implementing the theory. The fact that the predictions of the Yilmaz theory are unique may be difficult to understand by a physicist who is accustomed to the arbitrary adjustable parameters of the Einstein theory.

How Yilmaz Derived His Theory

Let us examine the steps by which Yilmaz derived his gravitational theory. As shown in Chapter 7, Einstein proved from his accelerating elevator that gravity produces a redshift corresponding to a wavelength ratio λ'/λ of approximately $(1 + \phi)$. Yilmaz implemented this analysis exactly, using the equations of special relativity, and found that λ'/λ is exactly equal to e^ϕ. Table 7-1 showed that λ'/λ is equal to $1/\sqrt{[g_{00}]}$, and so g_{00} must be equal to $1/(e^\phi)^2$, which is $e^{-2\phi}$.

Yilmaz postulated that the speed of light measured locally does not vary with direction. To satisfy this requirement, he proved that g_{11}, g_{22}, g_{33} must be equal in rectangular coordinates, and the $(g_{00}g_{11})$ product must be equal to -1. Hence g_{11}, g_{22}, g_{33} must be equal to $-1/g_{00}$ in rectangular coordinates.

With these remarkably simple calculations, Yilmaz derived all of the metric tensor elements for the static solution of his theory of gravity. From these metric tensor elements he calculated the corresponding gravitational field equation for his theory. The detailed steps of this analysis are given in Chapter 3, Section 3.5 of *Addendum* [2].

It took many years of difficult research for Yilmaz to extend his static solution to derive his general time-varying theory. This extension was essential to give the Yilmaz theory a solid theoretical foundation. Nevertheless, the simple static solution is more than adequate for nearly all practical applications.

Chapter 10

Weaknesses of the Einstein Theory

The Einstein theory has serious weaknesses, but these were not apparent during Einstein's lifetime, primarily because of the great mathematical complexity of the theory. Although the Schwartzschild limit was an annoying problem, Einstein considered this limit to be physically irrelevant, because it requires a mass-to-radius ratio that is one-quarter million times greater than the maximum value experienced within our solar system.

In 1939, Oppenheimer and Snyder predicted that a star must collapse to form a "black-hole" singularity if the mass-to-radius ratio of the star exceeds the Schwartzschild limit. Einstein strongly opposed this concept, because it drastically violates our laws of physics. He convinced himself that the Einstein theory does not predict a singularity condition. No one could refute Einstein until computers became widely available in the 1960's, about a decade after his death.

The physically impossible Schwartzschild black-hole singularity indicates that the Einstein theory has a mathematical weakness. In this chapter we will examine other weaknesses of the Einstein theory that have become apparent since Einstein's death. We saw in the last chapter that the Einstein theory lacks a tensor to characterize the energy in the gravitational field. This lack is a fundamental cause of limitations in the Einstein theory.

Does Not Achieve a Two-Body Solution

Professor Carroll O. Alley of the University of Maryland is one of the very few experts in general relativity theory who has applied his knowledge to practical applications. Prof. Alley has supervised several experiments to test the validity of predictions derived from the Einstein

10. Weaknesses of Einstein Theory 129

theory. This has included laser measurements with retro-reflectors on the moon that have allowed the distance to the moon to be measured with a laser beam to an accuracy of 3 centimeters. Another set of experiments measured the relativistic time delay in an atomic clock carried in an aircraft under several flight profiles. Prof. Alley is also intimately involved in applying general relativity corrections to the Geophysical Positioning System (GPS). The GPS is an array of satellites operated by the U. S. Air Force to provide accurate position coordinates over the world for military and civilian navigation.

Professor Alley became impressed with the Yilmaz theory, and has cooperated with Prof. Yilmaz in technical papers. Prof. Alley made an important contribution to this issue with his proof that the Einstein theory cannot achieve a multi-body solution. Let us examine this astonishing claim.

The Schwartzschild analysis of the Einstein theory was only a single-body solution. It merely considered the gravitational field produced by a single star. When this analysis was applied to calculate the relativistic advance of the Mercury orbit, the analysis considered only a test mass in the orbit of Mercury, which had no effect on the gravitational field.

Einstein recognized that this was only a single-body solution. He and other scientists were content with this limited analysis because it was not practical during Einstein's lifetime to achieve a multi-body solution with the very complicated tensor equations of the Einstein theory. A multi-body solution of the Einstein theory would require a non-diagonal metric tensor, which would result in millions of terms in the analysis.

Einstein assumed that his theory could yield a multi-body solution, but the following discussion shows that it cannot. More precisely, *the Einstein theory cannot yield an interactive multi-body solution.*

Many computer studies using the Einstein theory have appeared to achieve multi-body solutions. However these studies employ artifices that help to make the iterative computer program converge to a solution. These artifices are inserting results into the solutions that are not actually coming from the Einstein gravitational field equation.

Prof. Alley has avoided this problem by considering a simple physical model that can be solved analytically by the Einstein theory. He calculated the gravitational attraction between a parallel pair of infinite slabs of matter. He found that the Einstein theory predicts that there is no gravitational attraction between the two slabs. [16]

This configuration is physically similar to models used in electronics to calculate capacitance. By assuming that the dimensions of the slabs are infinite relative to the separation between the slabs, one can ignore edge effects. This results in a simple theoretical model to which one can apply the Einstein theory analytically. The analysis shows that the gravitational force between the two slabs is zero. This result conflicts with Newton's law of gravitational attraction and with experimental evidence.

The basic problem with the Einstein theory is that its gravitational field equation does not have a tensor to characterize the energy and stress of the gravitational field. In the vacuum of space, the energy-momentum tensor must be zero, and so all of the components of the Einstein curvature tensor must be zero. In the Alley analysis, there are no tensor components in the space between the two slabs to produce gravitational attraction between the slabs.

Einstein Curvature Tensor in a Vacuum

A fundamental problem with the Einstein theory is that the Einstein tensor G_μ^ν, which describes the curvature of space in the Einstein theory, is always zero in a vacuum. The reason for this is that the energy-momentum tensor T_μ^ν must be zero in a vacuum, and the Einstein gravitational field equation sets the Einstein tensor G_μ^ν proportional to T_μ^ν. Let us consider the implications of this property.

The Einstein tensor G_μ^ν is a minor modification of a related tensor called the Ricci tensor, denoted R_μ^ν. The Ricci tensor R_μ^ν is derived from a general tensor called the Riemann tensor. The Riemann tensor has 4 indices, and so has 4x4x4x4 elements, or 256 elements. The Ricci tensor is calculated from the Riemann tensor by a process called "contraction". Two of the indices of the Riemann tensor are set equal to one another, and the resultant expression is summed over the four values (0, 1, 2, 3) of these two indices, to obtain the Ricci tensor. Like the Einstein tensor, the Ricci tensor has two indices and hence has 4x4 or 16 elements.

In the Einstein theory, gravity is described as a curvature of space. The curvature of space is uniquely specified by the Riemann tensor. If space has no curvature (and hence no gravitational field), all elements of the Riemann tensor are zero. Likewise, if the Riemann tensor is identically zero, space has no curvature and no gravitational field. However, the Einstein gravitational field equation does not use the Riemann tensor. Instead it uses the Ricci tensor, or the closely related Einstein tensor, to describe the curvature of space.

This leads us to a strange observation. In the Einstein theory, the energy momentum tensor is zero in a vacuum, and so the Ricci and

10. Weaknesses of Einstein Theory

Einstein curvature tensors must always be identically zero in a vacuum. This means that in the Einstein theory a vacuum always acts approximately as if it has no curvature, and hence no gravitational field.

Of course it is the Riemann tensor that precisely describes the curvature of space. Even though the Ricci and Einstein tensors are identically zero in a vacuum, the Riemann tensor is not necessarily zero, and so a vacuum can theoretically still have curvature and a gravitational field. Nevertheless the Einstein equations do not directly recognize that a gravitational field is present in the vacuum. They use the Ricci tensor, not the Riemann tensor. Consequently in the Einstein theory a vacuum always acts *approximately* if it has no gravitational field. How then can the Einstein theory describe the effects of gravity?

The issue is confused by the fact that the Einstein theory appears to work. It yielded the Schwartzschild solution, which provided the basis for experimental tests to verify the Einstein theory. However that success disguised the fact that the theory can be applied reliably only to a single body having a weak gravitational field. The Einstein theory cannot characterize the gravitational interactions of multiple bodies, and it does not yield reliable predictions under intense gravitational fields, even in a single-body solution. The Schwartzschild singularity condition, which has led to the physically impossible black-hole concept, is a consequence of the failure of the Einstein theory to give physically meaningful results in an intense gravitational field.

Because of the great mathematical complexity of the tensor analysis on which the Einstein theory is based, this physical principle was not recognized by Einstein or by other scientists when general relativity was developed. Yet it lies at the heart of the contradictions that have evolved from the application of the Einstein theory.

The Yilmaz theory has a stress-energy tensor for the gravitational field, and so the Ricci and Einstein curvature tensors for the Yilmaz theory are usually not zero in a vacuum. This allows the theory to yield general multi-body solutions, and to give reliable predictions in an intense gravitational field.

The principles on which Einstein built his general theory of relativity were very sound, and represented a tremendous scientific advance. Einstein understood the desirability of having a tensor in his theory to characterize the gravitational field, but was unable to isolate a true tensor for this purpose. His theory appeared to work without it, and so he assumed that he had achieved a rigorous theory.

After Einstein's death, serious weaknesses of the Einstein theory have become apparent, which show that the theory does not provide a general rigorous solution. Since the Ricci and Einstein curvature tensors must be zero in a vacuum, the Einstein theory cannot describe the gravitational interactions of multiple bodies, and it cannot give reliable predictions in an intense gravitational field.

Conservation of Matter-Plus-Energy

In order to yield realistic predictions, a relativity theory must achieve conservation of matter-plus-energy. Since matter can be converted into energy, and vice-versa, it is the sum of matter-plus-energy that must be conserved. This conservation is achieved in a gravitational theory by placing appropriate constraints on the energy-momentum tensor.

Two different mathematical constraints on the energy-momentum tensor have been claimed to achieve conservation of matter-plus-energy. These can be expressed verbally as follows:

(a) The covariant derivative of the energy momentum tensor must be zero:
(b) The tensor-density derivative of the energy-momentum tensor must be zero.

We do not need to consider the actual mathematical formulas for these constraints. It is sufficient for us to recognize that the constraints are mathematically different. Those readers who desire further information should refer to the website *Addendum* [2] Section 3.6 of Chapter 3, and to Chapter 13 and Appendix J of *Universe* [1]. Appendix J explains the covariant derivative.

According to the "Bianchi identity", the covariant derivative of the Einstein tensor $G_\mu^{\ \nu}$ must always be zero. Since the energy-momentum tensor $T_\mu^{\ \nu}$ of the Einstein theory is proportional to the Einstein tensor $G_\mu^{\ \nu}$, the covariant derivative of the energy momentum tensor $T_\mu^{\ \nu}$ for the Einstein theory must be zero. Consequently Condition (a) is always satisfied by the Einstein theory. It is commonly believed that conservation of matter-plus-energy is assured if Condition (a) is satisfied.

However, Yilmaz has demonstrated that this conclusion is not true. This point is stated forcefully in the highly-respected book on relativity theory written by the Russian authors Landau and Lifshitz [17] (page 180). This reference shows that Condition (a) does not achieve conservation of matter-plus-energy. It demonstrates that Condition (b) must be satisfied to achieve this property.

Yilmaz has shown that Condition (b) is always satisfied by the Yilmaz theory. The energy-momentum tensor for the Yilmaz theory is constrained by the "Freud identity", which specifies that Condition (b) must be satisfied by the Yilmaz theory.

Condition (b) is mathematically inconsistent with Condition (a). Except in special cases, only one of these constraints can apply. Since the Einstein theory must satisfy Condition (a), it usually cannot satisfy Condition (b), and consequently the Einstein theory usually cannot achieve conservation of matter-plus-energy.

This discussion has shown that the energy-momentum tensor for the Yilmaz theory always achieves conservation of matter-plus-energy, but that for the Einstein theory generally does not. Therefore the energy-momentum tensors for the Einstein and Yilmaz theories are generally different, even when they are calculated from the same physical model.

Singularities Predicted by the Einstein Theory

The most obvious evidence that there is something fundamentally wrong with the Einstein theory are its predictions of singularities. Einstein insisted that, "Schwartzschild singularities do not exist in physical reality", because he recognized that the singularity associated with the black hole strongly violates our laws of physics. Einstein based his research on the principle that theory must agree with observation. During Einstein's lifetime, it was suspected that the Einstein theory predicts a singularity condition, but this suspicion was not proven.

After Albert Einstein's death, computer studies of the Einstein theory demonstrated that a very massive star must contract to form a black hole singularity. Big Bang theorists insist that the black hole singularity must be real, because its validity is proven by the Einstein theory. They fail to admit that this prediction strongly violates the philosophy that Einstein maintained throughout his lifetime. An obvious answer to this dilemma is to recognize that the Einstein gravitational field equation should be revised.

In Chapter 9 we saw that Einstein derived his gravitational field equation in a somewhat intuitive manner, and that various alternatives to this tensor equation have been proposed by scientists. Why do Big Bang theorists insist that the Einstein gravitational field equation must be infallible truth? Einstein is not here to give his opinion, but is difficult to believe that he could have accepted the physically impossible astronomical predictions that have been derived from his gravitational field equation since his death.

The singularities derived from the Einstein theory have resulted in many non-physical concepts. We observe in books, magazines, and television the astounding statements made by astronomers claiming that unbelievable processes are occurring within our universe, which radically violate our laws of physics. The January 2001 issue of *Scientific American* states on its cover: "Brave New Cosmos. Can the universe get any stranger? Oh, yes."

Because astronomy today lacks scientific objectivity, and has eliminated open scientific debate, it has descended into mythology. As Nobel laureate Hannes Alfven stated in a quote given in Chapter 2, "To most people it [has become] increasingly difficult to find any difference between science and science fiction."

Consistency with Quantum Mechanics

After presenting his general theory of relativity, Einstein did little with this theory. He devoted the rest of his life primarily to the task of developing a *unified field theory*, which would combine the principles of gravitational fields, electromagnetic fields, and atomic nuclear fields into a single theory. He never succeeded, although he struggled with this task until his last days. An important reason for his failure is that the singularities inherent in the Einstein gravitational field equation make it inconsistent with quantum mechanics.

Yilmaz has proven that his gravitational field equation is mathematically consistent with quantum mechanics. This finding opens great possibilities for relating relativity theory to quantum field theory, which may eventually lead to Einstein's elusive goal of a *unified field theory*.

Chapter 11

The Yilmaz Cosmology Model

The Yilmaz cosmology model is a simple application of the Yilmaz theory, which postulates that the universe has a constant average density of matter that extends to infinity and does not change with time. Since the Yilmaz gravitational theory yields a unique and rigorous solution, the predictions of this model should be taken seriously.

This model was presented briefly in the first paper on the Yilmaz gravitational theory, published in the *Physical Review* in 1958. Since that time, Prof. Yilmaz has ignored cosmological applications of his theory, because he came to realize that cosmology can be very speculative.

The author is exploring the cosmological implications of the powerful Yilmaz gravitational theory. Analyses were performed to extend the cosmology model that Yilmaz presented in his 1958 paper. This chapter summarizes the results of that analysis.

Description of Yilmaz Cosmology Model

Reduction of Speed of Light, Clock Rate, and Spatial Dimensions

We saw in Figs 7-5 and 7-6 of Chapter 7 that the gravitational field of a star causes the speed of light to decrease, it causes a spatial dimension to contract, and it causes a clock to run slower. The Yilmaz Cosmology Model predicts that the gravitational field produced by matter in the universe should cause similar effects as we look out into space. The results are shown in Figure 11-1. The solid curve shows how the speed of light appears to decrease with distance, and the dashed curve shows how a clock rate and a spatial dimension appear to decrease with distance.

136 *How Was Our Universe Created?*

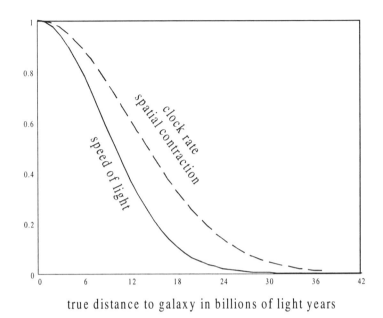

true distance to galaxy in billions of light years

Figure 11-1: Apparent speed of light (solid), clock rate (dashed), and spatial contraction (dashed), versus distance to a galaxy

These plots for the Yilmaz cosmology model assume that the Hubble constant H_0 is 25 km/sec per million light years. This Hubble constant corresponds to an apparent age of the universe T_0 of 12 billion years. Our analysis is based on the constant r_0, which is the distance that light travels during the apparent age T_0. Hence our value for r_0 is 12 billion light years.

Recent data suggests that the Hubble constant is probably close to 20 km/sec per million light years. For this case the apparent age of the universe would be 15 billion years, and the value for r_0 would be 15 billion light years. However, there is still appreciable uncertainty concerning the value for the Hubble constant, and so this book continues to assume a Hubble constant of 25 km/sec per million light years, which corresponds to an r_0 value of 12 billion light years.

The plot of spatial contraction given in Fig 11-1 shows that a dimension appears to contract with distance, and so the apparent distance to a galaxy is decreased. Figure 11-2 gives a plot of the apparent distance to a galaxy versus the true distance. Because of the strong contraction of spatial dimensions at great distances, the maximum

11. The Yilmaz Cosmology Model 137

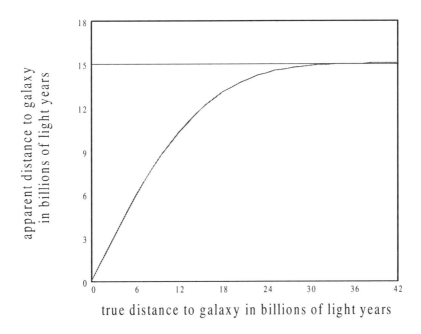

Figure 11-2: Apparent distance to galaxy verses true distance

apparent distance to a galaxy is finite, even though the model assumes that the true galaxy distance extends to infinity.

The maximum apparent distance to a galaxy is equal to $\sqrt{[\pi/2]}\, r_0$, where we have assumed r_0 to be 12 billion light years. For this value of r_0, the maximum apparent distance is 15.040 billion light years, or approximately 15 billion light years.

A distant galaxy appears to be closer than it actually is. Nevertheless, the apparent time for the light to reach us from the galaxy is not reduced, because the apparent speed of light and the apparent clock rate also decrease with distance. Combining the spatial contraction, speed of light, and clock rate, plotted in Figure 11-1 shows that the apparent time for light to travel between two distant points is the same as if there were no relativistic effects.

For example, consider a distance where the spatial contraction is ½. The relative speed of light is ¼ and the clock rate is ½. With distance reduced to ½, light should take twice as long to travel between two points if the speed of light is ¼. However, the apparent clock rate is cut in half, and so the apparent time for light to travel between the two points is the same as if there were no relativistic effects.

138 *How Was Our Universe Created?*

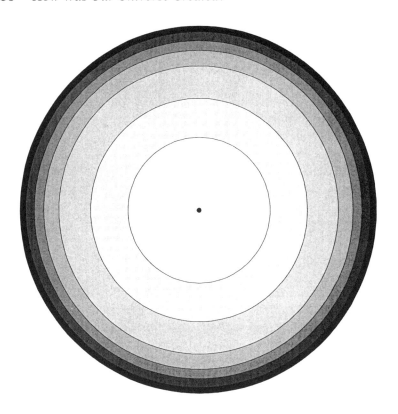

Figure 11-3: Apparent relative mass density of universe seen from earth; boundaries at density values of 1.5, 3, 10, 30, 100, 1000; minimum and maximum radii at 6 and 15 billion light years.

There is a very high contraction of the dimension of a galaxy that is close to the apparent limit of 15 billion light years. Consequently the apparent density of matter becomes extremely high as one approaches this limit at 15 billion light years. This effect is shown in Fig. 11-3.

Figure 11-3 shows how the density of the universe appears to vary with apparent distance from the earth. The darker the picture, the greater is the apparent mass density. The circles are contours of constant apparent mass density, and correspond to relative mass density values of 1.5, 3, 10, 30, 100, and 1000. The inner circle is at an apparent distance of about 6 billion light years, and the circumference is at an apparent distance of about 15 billion light years. The circumference theoretically corresponds to an infinite value of apparent mass density. However, light cannot reach us from galaxies beyond 45 billion light years of true distance. Hence the apparent mass density does not become infinite, although it can be very large.

We should realize that the "apparent" effects shown in Figs. 11-1 to 11.3 are real effects. As was explained in our discussion of special relativity, all reality is relative. There is no such thing as absolute reality. Any observation that we make from earth will display the "apparent" characteristics of the universe that we have examined.

Nevertheless we can still consider a "true" picture of the universe where galaxies are at a "true" distance from the earth. This "true" picture gives us a simple absolute model for understanding the universe.

The Hubble Expansion of the Universe

In general relativity theory, the geodesic equations are used to calculate the trajectory of a planet or any other body, or of a particle, or even of a light photon. To verify the Einstein theory, the geodesic equations were applied to the Schwartzschild solution to calculate the bending of a light ray when it passes close to the sun and the advance of the orbit of Mercury that is caused by relativistic effects.

When the geodesic equations are applied to the Yilmaz cosmology model, they show that a distant galaxy must recede at a velocity approximately proportional to its distance. In other words, the universe must expand just as Hubble observed. The results of this analysis are shown in Fig. 11-4. The solid curve shows the ratio of apparent receding velocity V_{ap} of a galaxy relative to the apparent speed of light c_{ap}.

The dashed line in Fig. 11-4 shows for comparison the ideal Hubble Law, in which galaxy velocity is exactly proportional to distance. According to the Hubble law, the velocity of a galaxy would reach the speed of light at a distance r_0 of 12 billion light years, and would exceed the speed of light at greater distances.

The solid curve shows that the receding velocity of a galaxy approximates the ideal Hubble Law plot out to a distance of about 5 billion light years. Thus the universe should expand approximately in accordance with the Hubble law within 5 billion light years. At much greater distances the apparent galaxy velocity gradually approaches the apparent speed of light, but never exactly reaches it.

The expansion rate that is predicted by the Yilmaz cosmology model varies with the mass density of the universe. The greater the density of matter, the faster the universe should expand. For our assumed Hubble constant, 25 km/sec per million light years, the average density of matter is equivalent to 7.5 hydrogen atoms per cubic meter. For a Hubble

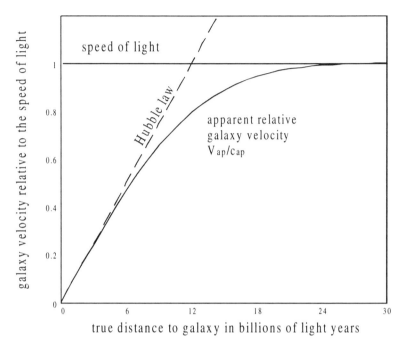

Figure 11-4: Apparent galaxy velocity relative to apparent speed of light, compared with Hubble law.

constant of 20 km/sec per million light years, the average density of matter would be 4.8 hydrogen atoms per cubic meter.

The solid curve in Fig. 11-4 represents the ratio V_{ap}/c_{ap}, which is the apparent receding velocity divided by the apparent speed of light. It is this apparent velocity ratio that astronomers calculate from the Doppler wavelength shift of galaxy spectra. They assume that they are measuring the true galaxy velocity relative to the true speed of light, but they are only measuring the apparent value of this ratio.

To see how the universe is actually expanding, we need the ratio V/c, which is the true galaxy velocity V divided by the speed of light c measured on earth. The spatial contraction plot in Fig. 11-1 is equal to the velocity ratio V_{ap}/V, and Fig. 11-1 also gives the relative speed of light c_{ap}/c. The following plots are combined to obtain the V/c ratio: the plot of V_{ap}/c_{ap} in Fig. 11-4, the speed of light plot of c_{ap}/c in Fig. 11-1, and the spatial contraction plot in Fig. 11-1, which is equal to V_{ap}/V. The resultant plot of V/c is shown in Fig. 11-5.

11. The Yilmaz Cosmology Model   141

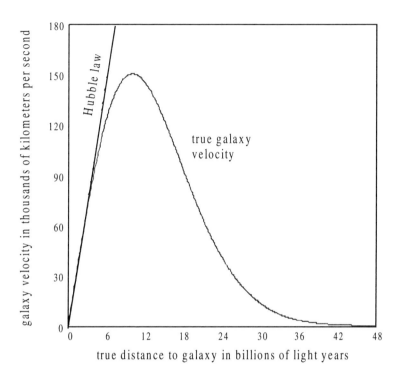

Figure 11-5: True galaxy velocity expressed in thousands of kilometers per second compared with Hubble law.

In Fig. 11-5, the speed of light c was set equal to 300,000 km/sec to give the true receding velocity V of a galaxy expressed in absolute terms. This shows that the true velocity V of a galaxy reaches a maximum value of 150,000 km/sec at a true distance r of 10 billion light years. This maximum value is half the speed of light c measured on earth.

At greater distances the true galaxy velocity decreases, and becomes very small at large true distances. For example, at a true distance of 36 billion light years, the true galaxy velocity is only 3000 km/sec, which is one percent of the speed of light measured on earth. At a true distance of 48 billion light years, the true galaxy velocity is only 100 km/sec.

Thus Fig. 11-5 shows that over very large distances the universe does not expand. It demonstrates that the Hubble expansion is a local relativistic distortion of space, not a general expansion of the universe.

How Can Gravity Make the Universe Expand?

The expansion of the universe displayed in Figs. 11-4 and 11-5 is a consequence of the Yilmaz gravitational theory. This indicates that the Hubble expansion of the universe is caused by gravity. "How can this be?", you ask, "How can gravity, which always causes masses to attract one another, force the universe to expand?"

Part of the answer is that the Hubble expansion is a local effect. Over very large distances the universe does not expand. Yet we are still faced with the question, "How can gravity cause the universe to expand locally?"

The rigorous mathematical answer to this question is that the geodesic equations, which characterize the effect of gravity, show that the universe must expand. This issue in discussed in Appendix C of *Universe* [1] and is proven mathematically in Chapter 4 of the website *Addendum* [2]. Nevertheless we would like an intuitive answer that makes sense physically.

Let us consider the following intuitive explanation. As was shown in Fig. 11-3, the whole universe appears to be compressed within a sphere having a radius of about 15 billion light years. The figure shows that the apparent density of matter is extremely high near the periphery of this sphere. We can assume that this outer shell of high-density matter is exerting gravitational force on the matter in the center of the sphere. This gravitational force would pull matter toward the periphery, and thereby produce the Hubble expansion. Thus *gravitational attraction* may be causing the *local expansion of the universe.*

However, this simple intuitive explanation is not consistent with gravitational force as described by Newton. Suppose that the universe is modeled as a thin spherical shell, in which matter is evenly distributed over the shell. If Newton's law of gravity is applied to a test mass placed inside the shell, the gravitational forces on the test mass are exactly cancelled. There is no net force attracting a test mass toward the spherical shell.

On the other hand, Newton's laws are only approximated in our relativistic model of the universe. If the gravitational forces inside the shell do not cancel exactly, there can be a net gravitational force attracting matter toward the mass in the shell. This postulate could then explain how the attractive effect of gravity can cause the universe to expand locally.

This gives a simple intuitive explanation of how gravity could cause the universe to expand. On the other hand, as was stated earlier, the rigorous explanation is that the universe expands locally because the geodesic equations show that it must.

Creation of Matter

The Yilmaz cosmology model assumes that the average density of matter does not change with time. In order to satisfy this requirement as the universe expands locally, the cosmology model requires that matter must be created to compensate for the expansion. Since matter and energy are equivalent, the creation of matter could be achieved by converting energy into matter. Therefore the Yilmaz cosmology model implies that energy is being transmitted across the universe to form matter that compensates for the Hubble expansion.

The rate of creation of matter that is required to compensate for the Hubble expansion is one hydrogen atom created per cubic meter every 500 million years. The rate of conversion of energy into mass to achieve this creation of matter is equal to 10 microwatts of power continually converted into matter within a volume the size of the earth.

Over very large distances the universe does not expand. Hence the total amount of matter and energy in the universe stays constant

Cosmic Microwave Background Radiation

Proponents of the Big Bang theory strongly acclaim the discovery of cosmic microwave background radiation as a milestone in validating the theory. In 1964 this radiation was first detected in a sensitive communication antenna at Bell Laboratories. Much more accurate measurements were obtained from the Cosmic Background Explorer (COBE) satellite in 1989. These experiments show that microwave background radiation is emanating uniformly from all directions, and has the spectrum and intensity that would be emitted from an ideal blackbody at a temperature of 2.73 degrees Kelvin.

Cosmic microwave background radiation was predicted by Big Bang theorists, who claimed it to be the cooled relic of radiation that was emitted from hot plasma 300,000 years after the Big Bang. However estimates of the blackbody temperature predicted by Big Bang theorists varied from 5 °K to 30 °K, and so the cosmic radiation was only predicted in a qualitative sense.

The Yilmaz cosmology model also predicts cosmic microwave radiation. Figure 11-4 shows that at very large distances the apparent galaxy velocity is very close to the apparent speed of light, and so the light radiated from distant galaxies should be Doppler shifted to very low frequencies. *Universe* [1] presents an analysis of this radiation in Appendix D, which is summarized as follows. The predicted radiation is equivalent to the emission from an ideal blackbody at a temperature between 2.1 °K and 3.4 °K. This is consistent with the 2.73 °K blackbody temperature measured by the COBE satellite.

The general spectrum of blackbody radiation is shown in Fig. 11-6. Diagram (a) shows the spectrum versus frequency and diagram (b) shows the spectrum versus wavelength. The wavelength scale is expressed in terms of the half-power wavelength λ_h, and the frequency scale is expressed in terms of the equivalent half-power frequency f_h, which is equal to c/λ_h. Half of the power falls at wavelengths greater than λ_h and half falls at wavelengths less than λ_h.

The temperature of a blackbody determines the actual frequencies that it radiates, but the shape of the spectrum is the same for all temperatures. The half-power wavelength λ_h is related as follows to the blackbody temperature T, expressed in degrees Kelvin:

λ_h = 4.107/T millimeter (mm)

The light radiated from our sun has a spectrum approximating that of an ideal blackbody at a temperature of 5770 °K. The corresponding value for the wavelength λ_h computed from this formula is 0.000712 mm, or 0.712 micrometers (millionths of a meter). Our analysis assumes that all stars in our universe have the same spectrum as our sun.

Because the light from a galaxy is Doppler shifted toward lower frequency, the equivalent blackbody temperature of the spectrum that is received from a galaxy decreases with distance to the galaxy. The solid curve in Fig. 11-7 shows the equivalent blackbody temperature of the Doppler-shifted spectrum that is received from a galaxy at a particular true distance r. This was calculated by applying the Doppler frequency shift formula to the data in Fig. 11-4, which shows the apparent galaxy velocity divided by the apparent speed of light for the Yilmaz cosmology model. It is this apparent velocity ratio that determines the Doppler frequency shift.

In Fig. 11-7 the equivalent blackbody temperature decreases from 5770 °K (the blackbody temperature of the sun) for a close galaxy down to 1°K for a galaxy at a true distance of 48 billion light years. (The dashed curve is an approximation, which can be ignored.)

11. The Yilmaz Cosmology Model   145

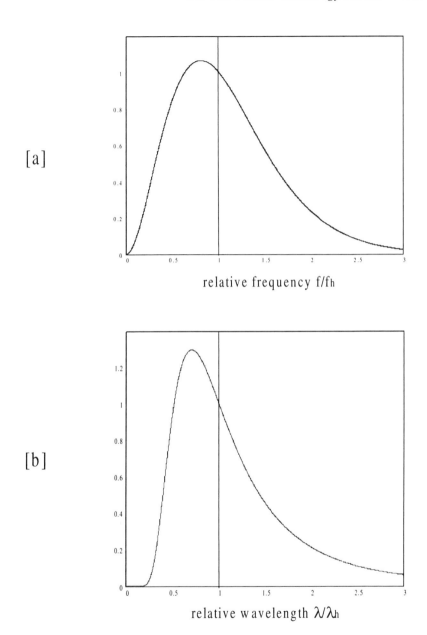

Figure 11-6: Normalized power spectra of a blackbody radiator; [a] versus frequency; [b] versus wavelength

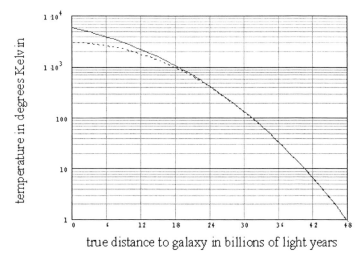

Figure 11-7: Blackbody temperature of radiation received from a distant galaxy versus the true distance r to the galaxy

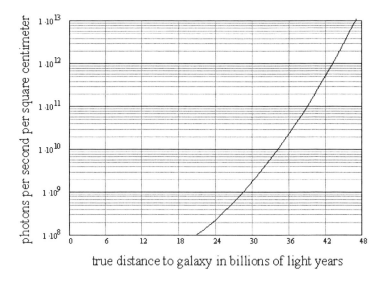

Figure 11-8: Photon rate per unit area for the cosmic radiation received from galaxies at a true distance r.

Figure 11-3 showed that the apparent density of matter becomes very high as the apparent distance to a galaxy approaches the limit of 15 billion light years. Because of this very high apparent density, the intensity of the radiation received from a galaxy located at a large true

11. The Yilmaz Cosmology Model 147

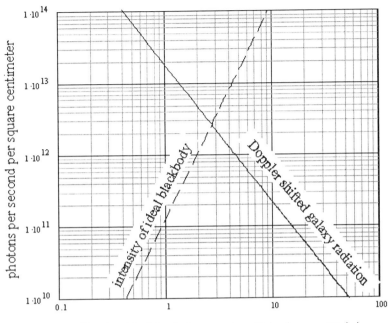

Figure 11-9: Photon rate intensity of cosmic radiation received from galaxies, versus effective blackbody temperature of radiation; received radiation (solid); ideal blackbody (dashed)

distance becomes very high. Figure 11-8 shows the photon rate that is received, per unit area of receiver surface, from galaxies at different values of true distance r.

When we combine the data in Figs. 11-7 and 11-8, we can plot the intensity of the received radiation versus the equivalent blackbody temperature of the spectrum. This gives the solid curve in Fig. 11-9. This shows the photon rate that is falling onto a square centimeter of receiver surface, expressed in terms of the equivalent blackbody temperature T of the received Doppler-shifted spectrum.

An ideal blackbody emits a photon rate that is proportional to the cube of the blackbody temperature. The dashed plot in Fig 11-9 shows the photon rate per unit area that is emitted from an ideal blackbody versus the temperature of the body.

For an ideal blackbody, the radiation is in thermal equilibrium with the molecules at the surface. We assume that this radiation level cannot be exceeded by cosmic radiation in space. If it were, the diffuse material in space should rapidly absorb the cosmic radiation. Therefore we conclude that the Doppler-shifted galaxy radiation indicated by the solid plot in Fig 11-9 cannot exceed the dashed plot. This indicates that the intersection point of the two plots should give the effective blackbody temperature of the received blackbody radiation. The figure shows that this calculated temperature is 2.7 °K. Because of approximations in the analysis, there can be an error in this result. We conservatively estimate that an exact computed temperature should fall within the range from 2.1 °K to 3.4 °K.

Figure 11-7 shows that galaxies producing blackbody radiation equivalent to a 2.7 °K temperature are at a distance of 45 billion light years. This indicates that we should receive from galaxies at a true distance of about 45 billion light years cosmic microwave radiation that corresponds to a blackbody at a temperature from 2.1 °K to 3.4 °K.

For comparison, the COBE satellite found that the received cosmic radiation is equivalent to the radiation from an ideal blackbody at a temperature of 2.73 °K. The COBE radiation is received with very high uniformity from all directions, which agrees with our analysis.

Density of Matter in the Universe

The Hubble expansion rate predicted by the Yilmaz cosmology model depends on the average density of matter in the universe. The density is equal to $(3H_0^2/8\pi G)$, where H_0 is the Hubble constant, and G is the gravitational constant of Newton's theory. For a Hubble constant of 20 km/sec per million light years, the average density of matter is equivalent to 4.8 hydrogen atoms per cubic meter.

Big Bang models of the universe define a critical mass density for the universe. According to the Big Bang theory, if the density of matter in the universe is less than the critical density, the universe will expand forever; and if the density of the universe is greater than the critical mass density, the universe will eventually collapse. The critical mass density for the Big Bang theory has the same value $(3H_0^2/8\pi G)$ as the mass density that is required by the Yilmaz theory. This shows that, in order to be valid, the Yilmaz cosmology model requires a density of matter that is equal to the critical mass density for the Big Bang theory.

The density of matter in the universe that is associated with the stars that we can see, called luminous matter, is much less than the critical mass density. However, there is much more dark matter in the universe (that we cannot see) than there is luminous matter. For example, the gravitational effects of dark matter in our Milky Way galaxy can be observed. There must be much more dark matter than luminous matter for the various parts of the Milky Way galaxy to rotate in the manner that is measured.

Big Bang theorists are searching hard for dark matter, because their theories yield a very low value for the age of the universe unless the density of matter is close to critical. However, they have not found sufficient dark matter in the universe to achieve critical mass density.

On the other hand, a fundamental mistake has been made in the Big Bang search for dark matter. Astronomers have used the radiation from quasars to estimate the density of intergalactic gas. Since they assume that quasars are at enormous distances, they have concluded that the density of intergalactic gas must be extremely small. Since quasars are very much closer, these estimates of intergalactic gas are not meaningful. Besides, most of the gas in space is probably molecular hydrogen (H_2), rather than atomic hydrogen (H), and molecular hydrogen is very difficult to detect.

It is probable that there is much more gas in the enormous spaces between galaxies than astronomers believe. This effect could explain the missing dark matter.

Uniqueness of Cosmology Model Predictions

The Yilmaz gravitational theory gives a rigorous and unique relativistic specification of the effects of gravity. The Yilmaz cosmology model applies this gravitational theory to a simple physical model of the universe. Since the Yilmaz theory is rigorous, and its predictions are unique, the description of the universe that has resulted from the simple Yilmaz cosmology model should be taken seriously.

For scientists accustomed to the Einstein theory, these principles may be difficult to understand. Chapter 9 showed that the Einstein theory yields multiple, contradictory solutions. It can be hard for these scientists to realize that the predictions of the Yilmaz theory are unique, and so are not the result of arbitrary hypotheses made by the individual who is applying the Yilmaz theory.

Chapter 12

Conclusions

The Implications of the Yilmaz Cosmology Model

Our Picture of the Universe

Let us explore the picture of the universe that has been predicted by the Yilmaz Cosmology Model. It yields a radically new and different concept of the universe, but is it correct? One thing is clear. There is no other cosmological theory that can withstand scientific cross-examination. Therefore a fundamentally different approach such as this ought to be seriously considered.

We start by returning to the principles of the Steady-State Universe theory, which assumes that the age of the universe is infinite. The universe has always appeared to be approximately like we see it today, and it always looks roughly the same from every point of the universe. As the universe expands, matter is being created to compensate for the expansion. This matter forms new stars and galaxies, and so the universe continually changes in detail. We accept these basic principles of the Steady-State Universe theory. However, the Yilmaz cosmology model does not suffer from the weaknesses that caused the original Steady-State Universe theory to be discarded.

We postulate that diffuse matter is continually being created throughout the universe. Matter is probably created one atom at a time in the form of the hydrogen atom, which has a single proton and an electron. This creation is achieved by the conversion of energy into matter, and so is consistent with our laws of physics. The energy includes electromagnetic waves and gravitational waves that are radiated from stars throughout the universe.

How is this conversion of energy into matter achieved? Although we

12. Conclusions

do not know this answer, it is clear that our postulate does not violate any law of physics.

The required rate of creation of matter is equivalent to one hydrogen atom created within a volume of one cubic kilometer every six months. One hydrogen atom is created every half-million years within a cube that is 10 meters on a side, which has the volume of a medium-size house. This creation of matter is achieved by the conversion of energy into matter. The rate of energy utilization is equivalent to the continual conversion of 10 microwatts of power into matter within a volume the size of the earth.

Because of gravitational forces, the diffuse matter that is continually created throughout the universe eventually congregates into clouds of hydrogen gas. A diffuse hydrogen cloud forms a galaxy, and bits of the galaxy coalesce to form individual stars.

As Lerner explains in his book, *The Big Bang Never Happened* [4], Nobel laureate Hannes Alfven has shown that plasma electric currents have greatly influenced the development of galaxies and stars. Many of the atoms of hydrogen gas in space are ionized, and the electrons can move freely. These moving electrons form electric currents that generate magnetic fields. The current flowing through a square meter of area is extremely small, but the total current flowing through the effective area associated with an individual star (several square light years) is huge. The magnetic fields produced by these currents generate vortices that cause stars and galaxies to rotate. These magnetic fields greatly affect the processes by which a galaxy gas cloud separates into stars and a solar system is formed around a star.

A bit of the galaxy gas cloud condenses to form a star. Energy released by gravitational collapse heats the hydrogen gas until the temperature is adequate to achieve nuclear fusion, in which the hydrogen is converted into helium. The energy from the fusion generates pressure that stops further gravitational collapse. A star is formed that can radiate energy for billions of years.

The rate at which a star generates energy depends on its size. The more massive a star, the faster its nuclear fusion operates, and the shorter is the life of the star. Most stars have a life cycle that is similar to our sun, but the rate at which the star progresses through its life cycle varies greatly with its mass.

Our sun is 5 billion years old. It is fueled by nuclear fusion, in which hydrogen is converted into helium, and eventually the helium will be converted into carbon. In 6 billion years our sun will have used up its available nuclear fuel, and will begin to shrink. Our sun will become a white dwarf, glowing white hot from energy released by gravity, until it shrinks to the size of the earth. When it reaches this size, the sun can shrink no more, and will gradually fade to become a black dwarf star, the dead ember of a once brilliant star. This basic life pattern is followed by 99.9 percent of all stars, but the rate at which a star ages depends on its mass.

One star in 1000 has sufficient mass to follow a radically different life pattern. Like the sun, this very massive star converts its hydrogen into helium and then into carbon. When this reaction is complete, the star has sufficient mass to generate the temperature and pressure required to achieve additional nuclear reactions, which form the elements oxygen, silicon, neon, nickel, cobalt, and iron. Then the nuclear fusion stops, because the generation of heavier elements absorbs energy; it does not release energy. The star suddenly suffers catastrophic collapse, and explodes as a supernova. For a brief period the supernova shines with the brilliance of millions of suns.

When this star collapses, electrons are forced into protons to form neutrons. This action releases neutrinos, which blow the star apart as a supernova. The neutrinos react with the atoms of the star to form elements heavier than iron. The elements generated earlier within the star and during the supernova explosion are scattered throughout the galaxy as dust particles. These dust particles are gathered by other stars to form the material from which solid planets like our earth are made. The matter that remains inside a supernova leaves a neutron star, which is 200 million times denser than the highly dense white dwarf.

We postulate that the light, heat, and other electromagnetic energy that is radiated from each star is transmitted across the universe, and is converted into diffuse matter in space. This matter in turn condenses to form new stars and galaxies, and the process continues indefinitely. However, this postulate alone could not result in a steady-state universe having infinite age. The universe would gradually deteriorate into black dwarf and neutron stars, and so would eventually die. There must be a process that converts the mass of these dead stars into energy.

The Einstein and Yilmaz theories both predict gravitational waves, and scientists are performing elaborate experiments that attempt to measure them. We postulate that the mass of a black dwarf or a neutron star is gradually converted into gravitational waves that are radiated. By this means, the matter within these dead stars would be gradually converted into energy.

Therefore we postulate that energy is radiated from stars in the form of electromagnetic waves and gravitational waves. This energy is converted into diffuse matter. It is possible that the creation of matter may result from an interaction between electromagnetic and gravitational waves. Although we do not understand how the energy is transformed into matter, our postulate is consistent with the laws of physics. In contrast, the Big Bang theory makes the physically impossible postulate that all of the matter and energy of the universe was created out of nothing, at the instant of the Big Bang.

Our cosmology model envisions a universe of infinite age. New stars in our universe are steadily being formed, as energy that is radiated from stars is converted into diffuse matter in space, which in turn condenses to form new stars and galaxies. Because of the continual creation of matter, the universe is always changing, and so our universe stays eternally young although it is infinitely old.

The Second Law of Thermodynamics

Scientists have often predicted that our universe must eventually run out of available energy and so must die. This conclusion is based on the *Second Law of Thermodynamics*. Let us examine this concept.

The US Patent Office does not prohibit a patent on a perpetual motion machine. It merely requires that one must submit a working model of the invention before it can be patented. There are two types of perpetual motion machines, which violate the two separate laws of thermodynamics. These laws of thermodynamics can be expressed in simple terms as follows:

> *First Law:* Energy can neither be created nor destroyed. Since matter and energy can be converted into one another, this is expressed in a more general form as: *The sum of matter-plus-energy can neither be created nor destroyed.*

> *Second Law: The availability of energy must continually decrease.*

The Second Law is rigorously expressed in terms of a concept called *entropy*, which roughly means the *degree of disorder*. The law states that *entropy* (the degree of disorder) must continually increase. However, the above definition gives a simpler interpretation of the law.

The following is an explanation of the *Second Law*. Consider a train locomotive driven by a steam engine. Water in the boiler is heated to generate steam, which pushes against the pistons to drive the wheels of the locomotive. The steam in the pistons is exhausted into the atmosphere. It can be shown that the energy that can be extracted from the heated steam depends on the difference in temperature between the steam in the boiler and the temperature of the air into which the steam is exhausted.

The gasoline engine in an automobile works under a similar principle. Its energy output depends on the difference of temperature between the exploding gasoline in the cylinders and the temperature of the exhaust. Gasoline engines are run as hot as practical in order to increase their efficiency. Diesel engines achieve greater efficiency than regular gasoline engines because they operate at higher temperatures.

A motor can theoretically run from any source of heat, provided that there is a sink at a lower temperature into which the heat can be discharged. However, useful energy cannot be derived from a heat source without a temperature differential.

With the passing of time, our world tends to become more uniform. Temperature differentials decrease, and so energy becomes less available. Our earth is replenished by energy that is radiated from our sun, which is generated by nuclear fusion occurring within the sun. Eventually the sun will run out of nuclear fuel, and will fade to become a dark body. When that occurs, the available energy on the earth will drop to zero.

Thus, in accordance with the *Second Law of Thermodynamics*, the availability of energy within our universe seems to be continually decreasing. Although matter-plus-energy remains constant within our universe, the availability of energy should theoretically decrease until the universe dies. How then can we justify a universe with an infinite age?

The Big Bang theory postulates that an enormous amount of energy and matter was created suddenly out of nothing at the instant of the Big Bang. This theory violates both laws of thermodynamics at the moment of creation.

The Steady-State Universe theory postulates that matter is being created very slowly out of nothing to form diffuse matter in space. Like the Big Bang theory, this theory violates both laws of thermodynamics, but does so continuously in infinitesimal amounts.

The Yilmaz cosmology model maintains conservation of matter-plus-energy, and so it satisfies the First Law of Thermodynamics. But what about the Second Law?

This cosmology model postulates that energy that is radiated from stars forms diffuse matter in space, which coalesces to form new stars. These new stars radiate energy, which forms new diffuse matter, and the process continues indefinitely. Matter-plus-energy is being conserved, in accordance with the First Law of Thermodynamics. However, the availability of energy does not decrease with time, and so the Second Law of Thermodynamics is being violated when we consider the universe as a whole. How do we justify this?

This contradiction is explained by the principle of *Relativity*, which allows the universe as a whole to violate the *Second Law of Thermodynamics*. Energy is being radiated from distant regions of the universe to generate the matter that compensates for the Hubble expansion. Since reality is relative, relativistic effects allow the *Second Law* to be violated by the universe as a whole, even though this law is satisfied locally.

In Chapter 3 we discussed the cosmological theory of Paul Marmet, which postulates that the Hubble redshift is an apparent effect, caused by photon collisions with hydrogen atoms. Marmet postulates that the Hubble redshift does not represent an actual expansion of the universe. We have rejected this theory because it requires a density of matter that it far in excess of observational evidence. However, a more basic limitation of this theory is that it does not account for the Second Law of Thermodynamics. If we assume the Marmet cosmology theory, the Second Law would eventually cause the universe to run out of energy and die. If the universe must eventually die, it must have had a beginning. What was that beginning?

This discussion has shown that the Hubble expansion and the continual creation of matter predicted by the Yilmaz cosmology model are essential features if a universe is to endure indefinitely. These features allow the universe to change continuously and thereby to stay eternally young even though it is infinitely old. Relativistic processes compensate for the *Second Law of Thermodynamics*, and thereby keep the universe from running out of available energy.

The Size of the Universe

How large is our universe, and how much matter does it contain? The Big Bang theory assumes that the observable universe has a radius equal to the distance that light travels during the apparent age of the universe, which is 12 billion years, if we assume a Hubble constant of 25 km/sec per million light years. For this Hubble constant, the radius of the observable universe is 12 billion light years.

Universe [1] gave an estimate of the amount of matter that should exist within a universe that has a radius of 12 billion light years. The analysis indicates that this universe should contain about 7 billion galaxies and 40×10^{18} luminous stars comparable to our sun. Note that 10^{18} mens 1 followed by 18 zeros, which represents one billion, billion. Therefore 40×10^{18} can be expressed as 40 billion, billion stars. There appears to be 100 to 500 times as much dark matter as luminous matter. Consequently the total matter within the Big-Bang observable universe is equivalent to at least 4000 billion, billion stars like our sun.

Our analysis indicates that cosmic microwave blackbody radiation is the Doppler-shifted effect of optical radiation emitted from galaxies at a true distance of 45 billion light years. The results of this analysis agree closely with the measured COBE data. This indicates that the universe should extend uniformly to a distance of at least 45 billion light years. Light radiated from beyond 45 billion light years cannot reach us, because it is absorbed by diffuse matter in space. Therefore we can regard 45 billion years to be the radius of the observable universe according to the Yilmaz cosmology model. Galaxies at this 45 billion light-year limit of our universe cannot be observed individually. The radiation from these galaxies is smeared together to form the cosmic microwave radiation.

The volume of the "observable" universe according to the Yilmaz cosmology model exceeds the volume of the Big Bang universe by the factor $(45/12)^3$, which is approximately 50, where 45 and 12 are the assumed radii of these two universe models in billions of light years. If we multiply the estimates given above for the observable Big Bang universe by 50, we find that the observable universe according to the Yilmaz cosmology model should contain about 350 billion galaxies containing 2,000 billion, billion luminous stars equivalent to our sun. The Yilmaz cosmology model predicts that there is about 500 times as much dark matter as luminous matter. Hence the total matter within this universe, including dark matter, may contain as much as one million, billion, billion times the mass of our sun.

At the limit of the observable universe, the apparent receding velocity of a galaxy is very close to the apparent speed of light. However, the apparent speed of light at that distance is only 234 meters per second, and the apparent galaxy velocity is close to this value. The true galaxy velocity is about 1000 times greater than this; it is 280 km/sec.

At a true distance of 60 billion light years, which is beyond the observable limit of 45 billion light years, the true receding velocity of a galaxy should be a mere 1 km/sec, which is only 1/30 of the velocity of the earth around the sun.

We have no means of estimating how far the universe might extend beyond 45 billion light years. We might postulate that the universe folds back onto itself, and so is not truly infinite. It seems reasonable to guess that the universe might have an effective radius of about 60 billion light years. If one could travel 60 billion light years in any direction, one would reach the same point. With this postulate, the volume of the total universe is about twice the volume of the observable universe. This leads to the estimate that the total universe has 700 billion galaxies, containing 4,000 billion, billion luminous stars equivalent to our sun. The total matter of the universe is about two million, billion, billion times the mass of our sun.

This estimate yields a universe that is infinitely old, which is extremely large and might actually be infinite in size. It has always looked more or less like we see it today. Since it continually changes, it appears to be eternally young even though it is infinitely old. A reasonable estimate of the total matter of the universe is two million, billion, billion times the mass of our sun. From the point of view of a mere mortal, this number is so enormous it might as well be infinite.

Mythological Status of Astronomy Today

If you should ask an authority in the field of astronomy about the Yilmaz cosmology model, the authority would probably smile and say, "One surely cannot take this concept seriously. There is a large competent group of scientists with verified academic credentials who can give a valid explanation of cosmology. You should listen to these experts to learn the truth."

However, we have seen that the field of astronomy today has descended into mythology. Scientific debate has been eliminated, and without scientific debate there can be no true science. One must take the findings achieved by astronomy today with "a grain of salt". Scientific careers are controlled by economic influences that force astronomers and astrophysicists to pursue the accepted dogma. Let us review the major evidence for this unfortunate state of affairs.

The Plasma Research of Hannes Alfven and Eric Lerner

The book by Eric Lerner, called *The Big Bang Never Happened* [4], was stimulated by the inability of Nobel laureate Hannes Alfven and other plasma physicists to have their research findings accepted by astrophysical journals.

There is strong evidence that plasma electric currents have greatly influenced the evolution of our universe. The book by Lerner gives a clear description of the manner in which plasma electric currents have affected the development of our solar system, our Milky Way galaxy, and our whole universe.

Many of the hydrogen molecules in space are ionized, and so the electrons can float freely from atom to atom. These electrons form electric currents, which generate magnetic fields. Over the vast distances of space, these currents and magnetic fields are huge. Hannes Alfven has shown that plasma electric currents, with their magnetic fields, tend to become unstable, and to form patterns that twist around one another like the strands of a rope. These effects can be observed in the aurora borealis, also called the northern lights, and they are also seen in gaseous nebulae. Alfven has demonstrated similar results experimentally in his laboratory. In his book, Lerner [4] has done an excellent job of explaining these remarkable plasma phenomena.

An important characteristic of astronomy is that everything seems to rotate. Stars and planets rotate around their axes, planets rotate around the stars, moons rotate around their planets, and galaxies rotate with a pattern that may exceed one hundred thousand light-years. What produces this rotation? The rotations are probably generated by electric currents in the plasma.

Alfven has shown that plasma currents can explain the processes that were involved in the birth of our sun, and in the development of our solar system. The initial cloud of gas and dust that formed our sun with its solar system was originally rotating because of the rotation of our galaxy. This cloud had thousands of times the angular momentum that our sun has today. How did the sun lose its angular momentum as this cloud condensed to form a star?

As the cloud contracts it rotates more rapidly, like a figure skater who spins faster by pulling the arms close to the body. The magnetic field of this rapidly spinning star interacts with the magnetic field of the surrounding plasma medium, and momentum is transferred from the star to the surrounding medium.

Studies indicate that a disk of gas and dust should be rotating around the star when it is formed. Plasma currents in this disk augment the collection of its material into planets. Therefore plasma currents have greatly influenced the formation of the earth and the other planets in our solar system.

We saw in Chapter 2 that the galaxies within our universe are nearly all arranged into long, curling, spaghetti-like filaments that extend for hundreds of millions of light years. Plasma theory can explain how electric currents could cause the filament arrangements of galaxies.

Despite the undeniable influence that plasma currents must have in our universe, cosmologists have largely ignored plasma theory. As Lerner [4] explains, few have even bothered to read about it. P. James E. Peebles stated that Alfven's ideas are "just silly". His colleague at Princeton, Jeremiah Ostriker, commented, "There is no observational evidence that I know of that indicates electric and magnetic forces are important on cosmological scales."

Alfven, as well as lesser-known plasma physicists, have repeatedly had their papers rejected by astrophysical journals because they contradict Big Bang wisdom. Alfven commented, *"I think the Catholic Church was blamed too much for the case of Galileo — he was just a victim of peer review"*.

It is clear that the implications of plasma research in astronomy are enormous. Nevertheless they have been brushed aside because they do not agree with the thinking of the scientific elite that controls the field of astronomy today.

The book by Eric Lerner gives an excellent documentation of the great potential of plasma research in astronomy, and the evidence that this research is largely being ignored. He has also presented a detailed and clear discussion of the severe conflicts between the Big Bang theory and observational evidence. However, Lerner failed to mention the most serious miscarriage of scientific objectivity in astronomy, which involved the astronomical quasar observations of Halton Arp. Let us review this case, which we have already discussed.

Observations of Quasar Redshifts by Astronomer Halton Arp

Halton Arp received his PhD degree in 1953 and performed outstanding research at the Palomar and Mount Wilson Observatories until 1984. He received many awards for his research. He was President of the Astronomical Society of the Pacific from 1980 to 1983, and received awards from the American Astronomical Society, the American Association for the Advancement of Science, and the Alexander von Humbolt Senior Scientist Award.

In 1984, the committee that controls use of the Palomar and Mount Wilson Observatories refused to allow Arp to use these facilities, because they did not like his research, which yielded strong astronomical evidence that conflicted with Big Bang dogma. Halton Arp was forced to accept early retirement, and he moved to the Max Planck Institute for

Physics and Astrophysics in Munich, Germany to continue his career.

The author highly recommends that the reader obtain a copy of Halton Arp's latest book, *Seeing Red* [5]. The evidence documented in this book gives overwhelming proof that quasars are relatively close objects. This evidence was very strong when Arp was denied use of the Palomar and Mount Wilson Observatories, but today it is enormous. Nevertheless, as explained in Arp's book, astronomical authorities continue to shun Halton Arp and to ignore his scientific findings.

If the field of astronomy can ostracize an individual with the prestige of Halton Arp, and bury his evidence because they do not like its implications, what kind of message is given to younger scientists in the field? How can astronomy achieve results that have scientific validity?

We see the consequences of this philosophy in the wild claims that are being made by astronomers today. We are told that we should believe these claims because many scientists having outstanding scientific and academic credentials have endorsed them. We should not be disturbed if these astronomical truths violate our common sense.

We should have absolute faith in these proclamations. We should believe that black holes are actually being found, proving that a condition of infinite density of matter actually exists. We should believe that quasars are enormously energetic objects located billions of light years away. They are probably caused by the interaction of two physically impossible black holes. (Do not be confused by the irresponsible statements by Halton Arp, who claims that quasars are nearby objects.)

Above all, we should have complete faith in the Big Bang prediction that our enormous universe began 15 billion years ago as a singularity that was smaller than a dime.

Contradiction with the Philosophy of Albert Einstein

It is ironic that the Big-Bang establishment controlling astronomical research continually quotes the great wisdom of Albert Einstein to support their position. Nevertheless, the "singularity" concept that underpins modern cosmology grossly violates the scientific philosophy that Einstein maintained throughout his lifetime. Einstein never accepted the black-hole singularity as representing physical reality.

Big Bang research assumes the infallibility of the Einstein gravitational field equation. Nevertheless Big Bang theoreticians admit that this equation was derived by Einstein in an intuitive manner, and that various responsible scientists have proposed alternatives to it. If

Einstein had lived to experience the physically impossible predictions that have been derived from computer studies of his gravitational field equation, he certainly would have tried to modify his equation.

Countless Big Bang theoreticians have based their professional careers on sophisticated computer studies of the Einstein general theory of relativity. Some of these theoreticians have been proclaimed to be "geniuses following in the footsteps of Einstein". However, the intellect that is required to apply the equations of the Einstein theory in a difficult computer study is radically different from the genius that was displayed by Einstein in developing the principles of relativity.

Need for a Theory of Gravity that Works

To develop an analytical study of cosmology, one needs a theory of gravity. We have seen that the Einstein gravitational field equation has serious mathematical weaknesses, and so cannot yield reliable predictions in an intense gravitational field. Consequently, it cannot be used as a foundation for a cosmological theory.

Many different approaches to cosmology have been developed that apply the Einstein general theory of relativity. We have referred to the Big Bang theory as a single theory, but it actually is a whole family of cosmological theories following the Big Bang concept. All of these apply the Einstein gravitational field equation, and so all suffer the same weakness. They are fundamentally wrong, because the Einstein gravitational field equation is flawed.

The Quasi-Steady-State Cosmology (QSSC) theory, recently proposed by Hoyle, Burbidge, and Narlikar, has developed an elaborate hypothesis in order to avoid the physically impossible singularity of the Big Bang theory. However, the QSSC theory applies the flawed Einstein gravitational field equation, and so has the same basic weakness as the Big Bang theory. This weakness is shared by nearly all cosmology theories.

This book, and the related documents, demonstrate that the Yilmaz theory provides a rigorous specification of gravity that has eliminated the problems of the Einstein theory. Hence the powerful Yilmaz theory is a reliable foundation for studying cosmology. This is where any acceptable theory of cosmology must start.

The Yilmaz cosmology model is a simple application of the Yilmaz gravitational theory to cosmology. It yields some remarkable conclusions, but are they correct? We know that this cosmology theory is based on a rigorous gravitational theory, which provides unique predictions. It satisfies an essential requirement that the other cosmology theories lack.

Religious and Philosophical Implications of Our Picture of the Universe

The instant of creation represented by the Big-Bang singularity has often been related to the story of Creation given in the Bible. Many have claimed that the Big Bang theory strongly reinforces this Biblical story. Nevertheless, we can find a solid basis for supporting the Biblical story of Creation without assuming that the whole universe began with a Big Bang 15 billion years ago.

We know that our sun was created 5 billion years ago. The earth, which defines our world, was created as a molten mass 4.6 billion years ago. The earth cooled and solid land was created. Our oceans were created from condensed steam and water from meteorites. The photosynthesis from life containing chlorophyll created the oxygen in our atmosphere, which allows us to breathe. The scientific story of the creation of our earth and the life on earth is all that one needs to support the principles of the Biblical story of Creation.

The universe as a whole has always existed, but individual stars and galaxies are continually being created. Matter is created to compensate for the local expansion of the universe, and is derived from energy radiated from other parts of the universe. There is a continual transfer of energy across the universe to compensate for the Hubble expansion, but the total mass and energy of the universe stays constant.

How much matter is in the universe? One might guess that it is about 2 million, billion, billion times the mass of our sun, which is twice the estimated mass within the observable universe. This number is extremely large but is not infinite. The universe has always had the same amount of matter and energy, and the total does not change with time. Since matter and energy are equivalent, it is the sum of the two that remains constant.

Thus our picture portrays an extremely large universe having a constant size. We may never be able to determine whether the size of our universe is finite or infinite. The universe has always existed, appearing approximately like we see it today, and will always remain that way. Yet the universe is continually changing because of the conversion of energy into matter. This creation of matter is what keeps the universe forever young, although it is infinitely old.

12. Conclusions

This is our picture of the universe. It is not a product of speculation. It evolved quantitatively from the Yilmaz theory of gravitation, which has a sound mathematical foundation. I personally find this universe picture to be warmly consistent with my religious beliefs. Those who reject religion should also find it to be philosophically satisfying.

The picture envisions a universe and a Creator of infinite age. The universe and the Creator have always existed, and will always continue to exist. Our world, the earth, was created 4.6 billion years ago. That represents the moment of *Creation* as far as mankind is concerned. Over billions of years, life developed on earth to form the world that we know today.

BIBLIOGRAPHY

[1] Adrian Bjornson, *A Universe that We Can Believe*, Addison Press, Woburn, MA, 2000, ISBN 09703231-0-7.
[2] Adrian Bjornson, *Addendum to, "A Universe that We Can Believe"*, available at no cost on Internet Website www.olduniverse.com.
[3] Internet *Website: www.olduniverse.com.*
[4] Eric Lerner, *The Big Bang Never Happened*, Times Books div Random House, NY, 1991, ISBN 0-8129-1853-3.
[5] Halton C. Arp, *Seeing Red*, 11998, Aperion, Montreal, Quebec, ISBN 0-9683689-0-5. (available at Internet website *www.Amazon.com*)
[6] Halton C. Arp, *Quasars, Redshifts, and Controversies*, 1987, Interstellar Media, Berkeley, Calif, ISBN 0-941325-00-8.
[7] David Filkin, *Stephen Hawking's Universe, the Cosmos Explained*, 1997, Basic Books div Harper Collins, NY, ISBN 0-465-08199-1.
[8] Peebles, Schramm, Turner, and Kron, "The Evolution of the Universe", *Scientific American*, Oct. 1994, pps 53-65.
[9] J. R. Oppenheimer and H. Snyder, "On Continued Gravitational Contraction", *Physical Review*, Sept. 1939, vol 56, pp 455-459.
[10] Albert Einstein, "On a stationary system with spherical symmetry consisting of many gravitating masses", *Annals of Mathematics*, Oct. 1939, vol 40, No 4, pp 922-936 (see p. 936).
[11] Albrecht Folsing, *Albert Einstein, a Biography*, 1997, (transl. from German by Ewald Osers). Penguin Books, NY, ISBN 0-14-02.3719-4.
[12] Geoffrey Burbidge, "Why Only One Big Bang", *Scientific American*, February, 1992, page 120.
[13] Fred Hoyle, Geoffrey Burbidge, and Jayant Narlikar, *A Different Approach to Cosmology*, 2000, Cambridge U. Press, United Kingdom, ISBN *0-521-66223-0.*
[14] Donald Goldsmith, *The Astronomers*, 1991, St. Martin Press, NY, ISBN 0-312-05380-0.

[15] John A. Peacock, Cosmological Physics, 1999, Cambridge U. Press, United Kingdom, ISBN 0-521-42270-1.
[16] Carroll O. Alley, "The Yilmaz Theory of Gravity and its Compatibility with Quantum Theory", *Ann New York Acad Science*, 1995 vol 755, pp 464-477.
[17] L. D. Landau and E. M. Lifshitz, *The Classical Theory of Fields, vol. 2*, 1973, Butterworth and Heinemann, Oxford, England, (first Russian ed. 1951) ISBN 0-7506-2768-9.
[18] Paul Marmet, "A New Non-Doppler Redshift", presented in Internet website: www.newtonphysics.on.ca.
[19] Paul Marmet, "Discovery of H_2 in space explains dark matter and redshift", originally published in *21st Century Science and Technology*, spring 2000, presented in Internet website www.newtonphysics.on.ca.
[20] Huseyin Yilmaz, "New Theory of Gravitation", *Proc. 4th Marcel Grossman Meeting Gen. Relativity*, Remo Ruffini, ed, Rome Univ, Italy, June 1985.
[21] Jayant Narlikar and Halton Arp, "Flat Spacetime Cosmology: A Unified Framework for Extragalactic Redshifts", *Astrophysical Journal*, vol. 405, pp. 51-56, March 1993.
[22] Jesse L. Greenstein and Maarten Schmidt, "The Quasi-Stellar Radio Sources 3C 48 and 3C 273", *Astrophysical Journal*, vol. 140, No. 1, July. 1964, pp. 1-34.
[23] S. Chandrasekhar, "The Dynamic Instability of Gaseous Masses Approaching the Schwartzschild Limit in General Relativity", *Astrophysical Journal*, vol. 140, No. 2, 1964, pp. 417-433.

INDEX

Numbers in brackets [ ] are Bibliography references.

absolute differential calculus, 87
acceleration of gravity, 90
aether, 74-76
age of earth and sun, 2, 24, 45
age of stars, 24
age of universe
   apparent age, 15
   true age, 6, 24-26
Alfven, Hannes, 18, 24, 37, 56, 157-159
algae, 46
Alpher, Ralph, 21
Alley, Carroll O, 128-130, [16]
amphibian, 48-49
Andromeda M31 galaxy, 13, 57
angular momentum,
   see momentum, angular
apparent relativistic effects,
   see, relativistic effects
archea, 45-46
Arp, Halton, 29-32, 36-37, 58, 109, 116-118, 159-160, [5, 6, 21]

bacteria, 45-46
background microwave radiation,
   see cosmic background radiation
Bianchi identity, 132
big bang theory, 1-32
birds, 50, 52-53

blackbody radiation, 20-22
   ideal, 21
   intensity, 147
   spectrum, 144-145
   temperature, 144
blackbody radiation, *see also,* cosmic background radiation
black hole, 7-10, 40, 104-106, 109-113
   Einstein refutation of, 9-11
Bondi, Hermann, 17, 33-34
Brahe, Tycho, 61
Burbidge, Geoffrey R., 23-24, 31, 34-36, [12, 13]

calculus, invention of, 66-67
   (by Newton and Leibnitz)
Cartesian coordinates
   see coordinates, Cartesian
Cepheid variable stars, 13-14
Chandreskhar, S., 108, [23]
clock rate, change of,
   special relativity, 82-84
   general relativity, 89-93, 104-108
conservation of energy and matter, 132-133, 153-155
contravariant tensor, 101
coordinates
   Cartesian, 94
   latitude-longitude, 95-96
   rectangular, 93-97
   spherical, 96-97
   4-dimensional, space-time, 100-101

168 How Was Our Universe Created?

Copernicus, 60-61
cosmology theories,
   see universe, models of
cosmic background explorer satellite (COBE), 22, 43, 143-144, 148
cosmic background radiation, 20-22, 143-148
   see also, blackbody radiation
   big bang prediction of, 20-22
   Yilmaz model prediction, 43, 143-148
   Bell Labs measurement, 20
   COBE satellite measurement, 22, 43, 143-144, 148
covariance, principle of, 85-86, 125-126
covariant derivative, 132-133
covariant tensor 101
creation of matter, 44, 143, 153-155, 162
creation of earth, 45, 56
creation of life, 45-55
creation of solar system, 55-57
creation of sun, 55-57, 151
critical mass density, 148-149
curvature of space, 88, 130-132

dark matter, *see* matter, dark
density of universe,
   see universe, density
density of matter, *see* matter, density
derivative, covariant, 132-133
Descarte, Rene, 94
de Sitter, Willem, 16
diagonal tensor, 98
Dicke, Robert, 21
dinosaur, 51-53
dimensions, contraction of
   see spatial contraction
distance, astronomical measurement, 11-14
Doppler wavelength shift, 11-12
dwarf star,
   white, 44

earth, formation, 45, 56
Einstein, Albert, 2-4, 9-10, 15-16, 75-76, [10]
Einstein general relativity,
   see also, relativistic effects,
   arbitrariness of, 4, 126-127
   clock rate, 89-93, 106-108
   computer studies of, 17-20
   curvature of space, 88
   nonphysical predictions, 7-11, 17-20
   pseudo-tensor, 125-126
   single-body solution, 128-130
   spatial contraction, 106-108
   tensors in, 87-88, 93-102
   variation of speed of light, 104-106
   verification of, 2, 88-89
Einstein gravitational field equation,
   see gravitational field equation
Einstein photo-electric effect, 76
Einstein relativity theories
   special, 82-86
   general, 87-108, 119-122, 124-134
Einstein tensor, *see* tensor types
electron, 111
electromagnetic fields and waves, 70-74
electromagnetic field equation, 70-71, 85-86
energy-momentum tensor,
   see tensor types
energy-plus-matter conservation, 132-133, 153-155
energy-to-matter conversion, 37, 44, 143, 153-155, 162
equivalence,
   between energy and mass, 84
   between acceleration and gravity, 89-93
ether, *see* aether
eukaryote, 46
event horizon, 104-106
eye glasses, 60

Index

field equation, gravitational, 119-126
field equation, electromagnetic,
    see electromagnetic field equation
Filkin, David, 7, 19, [7]
fish, 48
Fischer, J. R., 25
FitzGerald, George, 75
flat space, 102
Folsing, 10, 19, [11]
Friedman, Alexander, 16
Fresnel, Augustin, 69-70
Freud identity, 133
fusion, nuclear, 56, 84-85, 110-111

galaxies,
    M31, M33: 13-14, 57
    M35, 1, 55
    Milky Way, 1, 55, 57
Galileo, 60-62
Gamow, George, 16-17, 20-21
Geller, Margaret, 25
geocentric theory, 60
geological period
    Cambrian, 47
    Carboniferous, 49
    Cretaceous, 51-52
    Edicara, 47
    Jurassic, 51
    Permian, 51
    Triassic, 51
Gold, Thomas, 17, 33-34
Goldsmith, 110, [14]
gravity, acceleration of, 90
gravity wave, 152-153
gravitational constant G
    103-104, 120
gravitational field equation
    Einstein, 2, 119-122, 124-127
    Yilmaz, 4, 122, 124-127
gravitational potential, 91-93, 122-124
gravitational potential tensor, 122-124
gravitational redshift, 89-93, 103-108,
    113-115

gravitational theories,
    Einstein (general relativity)
        39-42
    Hoyle-Narlikar, 36, 39
    Yilmaz, 40-42
gravity, relativistic effects
    clock rate, 89-93, 102-108, 135-139
    Hubble expansion, 139-143
    spatial contraction, 102-108, 135-139
    speed of light, 102-108, 135-139
    wavelength, 89-93, 102-108
Greenstein, Jesse L., 114-115, [22]

Hawking, Stephen, 7, 19
heliocentric theory, 60
Herman, Robert, 21
Hertz, Heinrich, 71
Hooke, Robert, 64, 70
Hoyle, Fred, 17, 31, 33-37, 39, [13]
Hubble, Edwin, 1, 11
Hubble expansion
    concept, 11-14
    constant, 11-14
Huchra, John P., 25
Huygens, Christiaan, 68
humans, related species,
    australopithecus, 54
    homo erectus, 54
    homo habilis, 54
    homo sapiens, archaic, 54
    homo sapiens sapiens, 54
    Neanderthal man, 54
hydrogen-helium conversion,
    84-85

indices in tensors, 97-101
intergalactic gas, 149
isotropic Einstein solution, 102

Kepler, Johannes, 60-61

Landau, L. D., 132-133, [17]
Lemaitre, Georges-Henri, 16

Lerner, Eric J., [4]
  big bang criticism, 18, 24-29
  cosmology theories, 37-39
  plasma physics, 56, 157-159
Leavitt, Henrietta, 13
Leibnitz, G. Wilhelm, 66
Lifshitz, E. M., 132-133, [17]
life on earth, 45-55
life in universe, 57
light rays , bending, 88
light, speed of,
  measurement of, 74-81
  constancy of, 81-82
  variation with acceleration, 87
  variation with gravity, 87, 104-106
  value, 79-81
light, theory of, 62-66
  aether, concept of, 67-68, 74-75, 79-82, 85
  corpuscular, 59, 67-68
  electromagnetic, 70-74
  wave, 59, 67-70
Local group of galaxies, 87
Lorentz, Hendrick, 75
Lorentz transformation, 75, 82-83

Magellanic clouds, 13, 57
magnetic field, 56, 70-74, 115
magnetism, electro, 70-74
mammals, 50, 52-53
Marconi, Gugliemo, 71
Marmet, Paul, 37-38, 155, [18, 19]
mass
  equivalent to energy, 84-85
  normalized relativistic mass, 103-104
mass density of universe, 148-149
  critical mass density, 148-149
  Yilmaz prediction, 148-149
matrix, 98
matter, dark, 149
matter, density of, 148-149
matter and energy conservation, 132-133
matter-to-energy conversion, 84-85
Maxwell, J. Clerk, 70-74

Maxwell's electromagnetic field eqs., 70-74
measurement for star and galaxy
  distance, 12-14
  velocity, 11-12
Messier, Charles, 55
Mercury, orbit of
  relativistic effect, 88-89, 129, 131
  single-body solution, 129, 131
meteorite, 45, 162
metric tensor, 101-103, 123-124
Michelson-Morley experiment, 74
microwave background radiation, see cosmic background radiation
Milky Way, 1, 55, 57
momentum, angular,
  in solar system development, 56
multi-body solution
  of Einstein theory, 39, 128-130
myth and cosmology, 26-29, 157-161

Narlikar, Jayant V, 31, 35-37, 39, [13, 21]
nebula, 1, 113-115
neutrino, 111
neutron, 111
neutron star, 111-112
Newton, Isaac,
  optics research, 62-66
  mechanics research, 66-67

observable universe
  big bang theory, 156
  Yilmaz cosmology model, 137-138, 148, 156
Oppenheimer, J. R., 9-10, [9]

parallax, 12-14
parsec, 14
Pauli exclusion principle, 111-112
Peacock, John A., 124-125, [15]
Peebles, P. J. E., 8, 21, [8]
Penrose, Roger, 7, 19
Penzias, Arno, 20-21, 34

photoelectric effect, 76
photosynthesis, 46
photon, 37-38
plants, terrestrial, 49-50
plasma physics, 24, 56, 151, 157-159
proton, 111
Proxima Centauri, 57
pseudo-tensor, 125-126
pterosaur, 52-53
pulsar, 112

quantum mechanics, 134
quasar, 109, 113-118
   gravitational redshift, 106-108,
      113-115
   spectral lines, 113-115

radiation, blackbody,
   see blackbody radiation
radio wave, 70-71
redshift
   velocity (Doppler), 11-12, 113-118
   gravitational, 102-108, 113-118
   Marmet effect, 37-38, 117-118
relativistic effects
   clock rate, 82-84, 89-93, 102-108
   spatial contraction, 822-84, 102-108
   speed of light, 102-108
   synchronization, 83-84
   wavelength, 89-93, 102-108
relativity theory
   special, 82-86
   general, 87-108
relativistic units, 103-104
reptiles, 50-53
   Diapsid, 50-52
   Synapsid, 50-52
Ricci, Gregorio, 87
Ricci tensor, see tensor types
Riemann, Bernhard, 87
Riemannian geometry, 76, 87
Riemann tensor, see tensor types

Schmidt, Maarten, 114-115, [22]
Schwartzschild, Karl, 8, 88
Schwartzschild-Einstein solution,
   88-89
   energy-momentum tensor, 119-132
   event horizon, 104-106
   exterior solution, 120-121
   interior solution, 120-121
   Schwartzschild radius, 104
   singularity, 7-10, 18, 104-108
Silk, Joseph, 27
singularity, 7-10, 18, 104-108
simultaneous events, 82-84
Snyder, H, 9, [9]
solar system, creation of, 55-57
sound,
   speed of, 77-79
   wave concept, 59, 72-77
space,
   curved, 87-88, 102, 130-132
   flat, 102
spatial contraction
   from gravity, 106-108
   from velocity, 82-84
   in Yilmaz model, 135-139
speed of light, see light, speed of
spectral lines,
   forbidden, 113-115
spectrum
   of blackbody, 143-143
   of quasar, 113-117
stars
   age of, 24
   distance measurement, 12-14
   velocity measurement, 11-14
stars, variable
   Cepheid, 13-14
   RR Lyrae, 14
   quasar variation, 113-115
steady-state universe theory
   see, universe. models of
stress-energy tensor, see tensor type

sun
　angular momentum, 56
　formation of, 56-57
　normalized mass, 103-104
　mass, 103
　radius, 104
supernova, 14, 111
synchronous clocks, 79-84

telescope types,
　field glass (Galileo), 60-61
　inverted image (Kepler), 61
　reflecting (Newton), 63
　achromatic lens, 62-63
temperature, blackbody,
　see blackbody temperature
tensor analysis, 87
tensor, basic concept, 97-102
tensor density, 132
tensor, elements of
　diagonal, 98
　trace of, 123
tensor forms of, 101
　(contravariant, covariant, mixed)
tensor, types of
　Einstein, 119-122, 130-131
　energy-momentum, 119-124,
　　132-133
　gravitational potential, 123-124
　metric, 101-104
　pseudo-tensor, 125-126
　relativistic, 100-101
　Riemann, 130-131
　Ricci, 130-131
　stress-energy tensor,
　　for matter,
　　for gravity field,
thermodynamics, laws of, 153-155
Tully, Brent, 25

units, relativistic, 103-104
universe age,
　apparent, 15
　true, 15, 33-34, 162
universe age dilemma, 24-26
universe, distances in, 11-14
universe, mass density of,
　see mass density
universe, models of, 7-44
　big bang, *see* big bang theory
　Marmet theory, 37-38, 155
　matter-antimatter theory, 37
　photon-hydrogen collisions,
　　see, Marmet theory
　quasi-steady-state, 35-36
　steady-state, 33-34, 39
　Yilmaz cosmology model, 42-44
universe radius, big bang
　observable, 136, 155-156
universe radius, Yilmaz
　apparent, 136-138
　observable, 156
universe, structure of, 24-26
　(filament and ribbon),
unified field theory, 134

vector, 93-97
velocity relativistic effect on
　clock rate, 82-83
　spatial contraction, 82-83
　simultaneity, 83
vertebrates, 47

wave
　electromagnetic, 70-74, 76
　light, 67-70, 76
　sound, 59, 76
　water, 59, 76
white dwarf, 110
Whirlpool galaxy, 1
Wilson, Robert, 20-21, 34

Yilmaz, Huseyin, 4, [20]
Yilmaz cosmology model. 5-6
  clock rate, 135-137
  creation of matter, 143
  cosmic background radiation, 143-148
  Hubble expansion, 5-6, 139-143
  density of universe, 148-149
  spatial contraction, 135-139
  speed of light, 135-137
  uniqueness, 149

Yilmaz gravity theory, 4-6
  derivation, 127
  gravitational field equation, 122-126
  stress-energy tensors,
    for matter, 122, 132-133
    for gravity field, 122, 130-132
  quantum mechanics, 134
  uniqueness of, 4, 126-127, 149
Yilmaz theory solutions,
  static, 122-124
  time-varying, 123-124
Young, Thomas, 68-70